T0135599

Combustion characteristics of turbo charged DISI-engines

Von der Fakultät für Umweltwissenschaften und Verfahrenstechnik der Brandenburgischen Technischen Universität Cottbus zur Erlangung des akademischen Grades eines Doktor-Ingenieurs (o. a. Titel nach § 1 (2)) genehmigte Dissertation

vorgelegt von

Diplom-Ingenieur

Henrik Hoffmeyer

aus Verden an der Aller.

Gutachter: Prof. Dr. Fabian Mauss
Gutachter: Prof. Dr. Bengt Johansson
Gutachter: Prof. Dr. Heinz Peter Berg
Tag der mündlichen Prüfung: 22.09.2011

Bibliografische Information der Deutschen Nationalbibliothek

Die Deutsche Nationalbibliothek verzeichnet diese Publikation in der
Deutschen Nationalbibliografie; detaillierte bibliografische Daten sind
im Internet über http://dnb.d-nb.de abrufbar.

ISBN 978-3-8325-3079-2

Logos Verlag Berlin GmbH
Comeniushof, Gubener Str. 47,
10243 Berlin
Tel.: +49 (0)30 42 85 10 90
Fax: +49 (0)30 42 85 10 92
INTERNET: http://www.logos-verlag.de

Affirmation

Hereby I declare on oath that I have written the present work without any inadmissible help of a third party and without any devices as mentioned. Ideas and notions from foreign sources are identified. Third parties do not have received any money equivalent benefits for any work, which is conjunct with this dissertation.

The present work has not been submitted in similar or equal form as a dissertation or published as a whole.

Hannover, März 2010

Henrik Hoffmeyer

Eidesstattliche Erklärung

Ich erkläre hiermit, dass ich die vorliegende Arbeit ohne unzulässige Hilfe Dritter und ohne Benutzung anderer als der angegebenen Hilfsmittel angefertigt habe. Die aus fremdem Quellen übernommenen Gedanken sind als solche kenntlich gemacht. Dritte haben von mir weder unmittelbar noch mittelbar geldwerte Leistungen für Arbeiten erhalten, die im Zusammenhang mit dem Inhalt der vorgelegten Arbeit stehen.

Die Arbeit wurde bisher weder im Inland noch im Ausland in gleicher oder ähnlicher Form als Dissertation eingereicht und bisher auch nicht als Ganzes veröffentlicht.

Hannover, März 2010

Henrik Hoffmeyer

Acknowledgement

The presented work has come into existence during my time as Phd student at the Volkswagen AG group research department for SI-engine technology. For giving me the chance to carry out this work, I like to thank Fred Thiele from Volkswagen AG. My primary gratitude goes to Prof. Dr. Fabian Mauss for mentoring my work and encouraging me. The few days spent at Cottbus have always been a pleasure. I thank Prof. Dr. Bengt Johansson as well as Prof. Dr. Heinz Peter Berg for their co-discussions.

My deepest gratitude goes to Dr. Kay Schintzel for mentoring my work at Volkswagen AG. His way of guiding, encouraging and motivating makes all this possible. Next to this, I deeply appreciate our discussions about engines and everything beyond.

Furthermore, I owe Dr. Emanuela Montefrancesco a debt of gratitude for her moral and CFD support. I thank Linda-Maria Beck for performing kinetical simulations, Florian Ziebart and Tim Thaler for their dedicated support on test bench measurements as well as Andreas Hildebrandt, Markus Welk and Heinz Ulrich Gieger for all the hours they spend on fitting the engine. I thank Andreas Börstler, Mike Sliwa and Christian Zabka for their mechanical and electrical work. I greatly acknowledge Dr. Gerhard Ohmstede, Michael Lauerhaas, Lars Kapitza and Mark Gotthardt for performing the high-pressure chamber measurements as well as Kerstin Schmidt and Thomas Rudolph for designing the endoscopic acces.

Special thanks go to Mario Campe, Dr. Andreas Grote, Uwe Reisch, Dr. Christian Jelitto as well as our doctoral regular's lunch table for their continous moral support and bearing my tempers. Not forgotten, I like to thank my office flat mate Dr. Jan Jakobs for his way of support and taking care of me.

Hannover, January 2012

Table of content

1 Introduction

From the invention of automobiles in the beginning of the 20th century, automobiles have been established as widespread mobility device as well as a symbol of individual mobility in the western hemisphere. Along with dispersion, a fast-paced technical evolution has taken place advanced by industrialization of western societies at the turn of century and technical achievement in all fields of automobile technology. Today, more than one hundred years later, the automobile has become a sophisticated article of daily use, which has to accomplish all requirements like easy handling, emotionality, economic efficieny as well as protection of natural resources. The exploitation of emerging markets and their mobility requirements demands for continous technical innovation and confronts the global society with new challenges.

In spite of progress in the development of alternative powertrain systems and energy sources, the internal combustion and all its derivates still are and will be the main powertrain for automobiles. The shortage of fossil fuels as well as the increasing stress of the environment, which needs to be limited by emission legislations, leads to increasing challenges for the development of future internal combustion engines. Enhanced by supra-national legislative bodies, OEMs are obliged to decrease the fleet averaged emission of carbon dioxide to 120 g/km in the New European Driving Cyle (NEDC) in 2015 and down to 95 g/km in 2020. To achieve this, intelligent electrification of the powertrain is essential just as combustion processes with increased customer relevant efficiency. In SI-engines, several approaches compete with each other like the controlled auto ignition (CAI or HCCI), throttle-free load control using variable valvetrains, stratified mixture formation with lean engine operation or highly turbo charged downsizing concepts [1]. All concepts use the gasoline direct injection to gain maximum benefit [2],[3].

1.1 The direct injection SI-engine

The gasoline direct injection is a well known technology to increase engine performance. The first known direct injection SI-engine in production has been the DB 601 aircraft engine developed by Daimler-Benz in the late 1930ies. In the 1950ies, Mercedes-Benz introduced the first direct injection SI engine in a production car in the MB 300 SL. In the 1970ies, Ford presented the first stratified combustion process in the Ford Proco. In the 1990ies, Mitsubishi reanimated the stratified combustion process and introduced the GDi® (Gasoline Direct injection) concepts into market. At the end of the 1990ies, the Volkswagen group presented the first generation of FSI® engines using stratified mixture formation, followed by Renault and Opel. Today, these engine concepts are out of production. The favoured concepts of air-/wall guided combustion systems were not able to approve their theoretical efficiency potential of stratification for the customers. Since 2003, the stoichiometric gasoline direct injection has been introduced into market by the Volkswagen group in naturally aspirated engines as well as in turbo charged engines (FSI®/TFSI®/TSI® technology) as well as by Toyota/Lexus, Alfa Romeo and Opel.

In 2006, DaimlerChrysler introduced the first spray-guided combustion process in the MB 350 CGI®. Using a centrally mounted piezo-injector and a NSC (Nitric Storage Catalyst) system for emission control, the engine operates in stratified mode up to a vehicle speed of 130 km/h [4]. In the beginning of 2007, BMW introduced a spray-guided combustion process in all naturally aspirated four- and six-cylinder engines. Emission control is achieved by the use of a NSC system [5]. Additonally in 2007, the first turbo charged direct injection engine with centrally mounted piezo injector was put into production by BMW [6].

Under neglection of exhaust gas emissions limits, an efficiency increase of fifteen to twenty percent can be realized comparing to port fuel injected gasoline engines using spray-guided stratified mixture formation [2],[7]-[9]. Limiting superior system relevant boundary conditions are primarily pollutant emissions like combustion products and the limited engine map area. To higher loads, spray-guided stratified engine operation is limited by the availability of excess air. To higher engine speeds, the carburetion becomes instabile due to shortening timescales. By the use of turbo charging, the engine map area with possible stratification can be extended to higher loads and speeds (compare Figure 1-1) [9],[10]. Under neglection of NO_x-emission limits, more than fifty percent of maximum brake torque (MBT) can be operated in stratified mode. Kuberczyk performed cycle process calculations and determined a decrease of fuel consumption of eighteen percent in the NEDC due to the combination of turbocharging and spray-guided stratification [11]. Because of neglection of NSC regeneration, the entire NEDC can be operated with stratified engine operation.

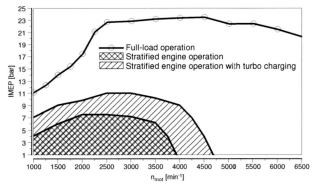

Figure 1-1: Engine map with full-load line and areas of possible stratified engine operation with and without turbo charging.

Nevertheless, due to carburetion characteristics, inevitable chemical losses and the combustion process, nitric oxide and soot emissions limit the theoretical efficiency [10],[12],[13]. Because of the overstoichiometrical mixture during stratified engine operation, nitric oxide emissions cannot be reduced sufficiently with a standard three-way catalyst. Thus, additional exhaust aftertreatment systems are necessary like the NSC catalyst or the SCR (SelectiveCatalyticReduction)-System implicating further limits. The NSC-Catalyst stores engine emitted NO_x emissions. After a certain time, it needs to be regenerated under enriched mixture conditions resulting in increasing fuel consumption. Hence, the use of a NSC-Catalyst has to be limited to lower NO_x emission mass flow limiting the engine map area with stratified engine operation. The SCR system allows the continous reduction of emitted NO_x emissions, but it requires an additional operating liquid - urea - for NO_x reduction. Today, SCR-systems are widely spread in heavy-duty applications, which use Ad-Blue® as additional operation liquid. Due to high development and production costs for SCR systems, todays direct injection SI-engines using stratified engine operation are supplemented with NSC-catalysts. The upcoming EU5+ and EU6 legislation limits particle emissions of direct injection SI-engines. Due to heterogenous mixture formation during stratified engine operation, the probability of soot emission towards higher engine loads, i.e. lower exhaust gas air-/fuel ratios, limits engine efficiency [14].

1.2 The turbo charged spray-guided combustion system

The investigated turbo charged spray-guided combustion process is realized in a four cylinder in-line engine. Based on a serial production engine, the cylinder head is modified in such way, that a piezo-driven fuel injector with an outward opening ring hole nozzle can be mounted centrally into the cylinder head. Additionally, the spark plug is moved towards the exhaust side and the inlet valve diameter is reduced for positioning the injector (compare Figure 1-2). The piston is a serial production piston with a slight central bowl allowing a geometrical compression ratio of 9.9:1. For improved full-load and an enlarged engine map area with stratified mode, the engine is turbo charged with a serial production waste gate turbo charger. Appendix Table A-7-1 shows the main data of the test engine.

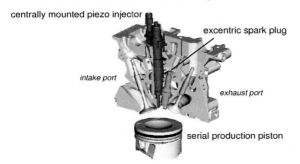

Figure 1-2: CAD model of the investigated turbo charged spray-guided combustion process.

For using the thermodynamic improvement of external recirculated exhaust gas (EGR), the engine is equipped with an EGR route. The high pressure EGR route consists of serial production engine water cooled EGR cooler and EGR valve. The EGR is extracted right in front of the turbo charger's turbine inlet and then led around the engine and fed into the intake path just behind the throttle in front of the intake manifold. The engine is equipped with a serial production exhaust gas aftertreatment system consisting of a three-way-catalyst and two exhaust sound absorbers generating real-life exhaust gas back pressure.

The engine is controlled using an open source research engine control unit (ECU) based on a standard ECU (BOSCH MED 7.1) and an additional component ECU for controlling the EGR valve. All application parameters are set individual. Fuel rail pressure is controlled using the serial production control algorithm of the Volkswagen

TSI® ECU family just as the valve timings and cam shaft adjustment are. A modified high-pressure fuel pump allows fuel rail pressure up to 210 bar. Using a hydraulic cam shaft phaser, the inlet cam shaft can be advanced about fourty crank angle degree to earlier timing.

1.3 Conceptual formulation

In differentiation to known publications, the stratified carburetion with high air excess is defined as spray-guided combustion process. Todays direct injection SI-engines with spray-guided combustion processes are limited by NO_x and soot emissions. NO_x emissions limitation is given by the available exhaust gas aftertreatment system. While NO_x formation and oxidation is sufficiently investigated and understood, soot emissions and the corresponding occurrence of non-premixed combustion phenomena in direct injection SI-engines are not completely understood. The numerical modeling of soot formation is being focussed in several research works and needs to be based on experimental combustion analysis. Below, the occurrence of non-premixed combustion phases and engine parameter influences on it during stratified engine operation are being investigated experimentally. Since soot formation is kinetically controlled, flame temperature is important information for successful modeling.

After a survey on fundamentals of heat release analysis of direct injection SI-engines and combustion, the model conception of the spray-guided combustion process is developed. Further on, fundamentals of soot formation, oxidation and engine emission are being discussed. For detailed combustion analysis, the application of optical measurement technologies is irreplaceable. Investigating non-premixed combustion phases, the 2-ColorRatioPyrometry (2-CRP) and its application on the test engine are specified. During experimental investigations, optical combustion characteristics of stratified combustion are discussed. Furthermore, two-dimensional temperature distributions of non-premixed flame are determined and discussed in respect to data of detailed heat release analysis. The influence of engine parameters like fuel rail pressure or external EGR rate on non-premixed flame temperature are displayed. To describe the influence of oxidization enhancing fuel components, ethanol enriched dual-component fuels are investigated and opposed to two multi-component fuels.

Under high- and full-load conditions, direct injection SI-engine efficiency is limited by engine knock. Using cooled external EGR, knock tendency can be reduced. After a survey on auto ignition fundamentals and the kinetical mechanism of engine knock, the effect of EGR on engine knock is investigated. Using detailed kinetical analysis simultaneously, the auto ignition influencing effect of EGR components is demonstrated. Based on this, EGR catalysis is developed using an EGR catalyst to reform EGR. Engine test bench investigations show the influence of reformed EGR on knock tendency, while detailed kinetical analyses deliver detailed information about the sensitivity of selective components.

2 Combustion and thermodynamics of turbo charged direct injection SI engines

As described before, the direct injecting gasoline engine shows a considerable potential for decreasing fuel consumption during part load operation due to the possibility of mixture stratification. In comparison to a port fuel injected gasoline engine with equal displacement volume, this potential can be reasoned when considering the chain of efficiency according to Weberbauer et al. [15]

$$\eta_e = \eta_{th} - \Delta\eta_{HC+CO} - \Delta\eta_{RoHR} - \Delta\eta_{Calorics} - \Delta\eta_{Wall\ Heat} - \Delta\eta_{Gas\ exchange} - \Delta\eta_m \qquad \text{[2-1]}$$

with

η_e		effective engine efficiency,
η_{th}		thermal efficiency,
$\Delta\eta_{HC+CO}$		losses due to incomplete combustion,
$\Delta\eta_{RoHR}$		losses due to finite heat release,
$\Delta\eta_{Calorics}$		losses due to changing gas properties,
$\Delta\eta_{Wall\ heat}$		losses due to wall heat transfer,
$\Delta\eta_{Gas\ exchange}$		losses due to gas exchange and
$\Delta\eta_m$		mechanical losses

by

- an increase of η_{th} due to a higher effective compression ratio. The internal carburetion cools the in-cylinder mixture and therewith lowers its knock tendency.
- an increase of $\eta_{Calorics}$ when using stratification due to air excess.
- a decrease of gas exchange losses ($\Delta\eta_{Gas\ exchange}$) due to lower throttled air suction when using stratification and
- a decrease of wall heat losses ($\Delta\eta_{Wall\ heat}$) due to the surrounding excess air.

In opposition to that, the increased exhaust gas mass flow, the increased friction losses due to higher pressure forces on the pistons during stratified operation and the mechanical generated fuel pressure reduce the thermodynamic potential [1].

2.1 Thermodynamic analysis of direct injection SI engines

For determining the influence of engine parameters on engine efficiency, the zero-dimensional thermodynamic analysis is a useful method. Based on the measured cylinder pressure for a working cycle, the thermodynamic process is calculated using semi-empirical models. Real thermodynamic cycle processes give the best approximation of the thermodynamical processes in a combustion engine. Here, the dependencies of the following parameters are respected:

- the composition of the in-cylinder mixture, consisting of fresh air, fuel and residual gas;
- the rate of heat release;
- the caloric gas properties depending on the actual temperature, pressure and mixture composition;
- the wall heat transfer and
- the cylinder mass, which can differ over time under consideration of leckage.

In this zero-dimensional approach, a homogenous mixture of the gas components in the control volume (equal to the combustion chamber) is assumed during all phases of the process. There are no local differences regarding pressure, temperature or cylinder mixture respected. For improved analysis during combustion, the combustion chamber can be separated in an unburned and a burned zone (two-zone-model). It is assumed, that these zones are separated by an infinite thin, massless flame front [16]. The thermodynamic analysis of combustion engines is based on three equations:

Equation of thermal state: $\qquad p \cdot V = m \cdot R \cdot T$ [2-2]

with: p actual pressure in the combustion chamber,

V actual volume in the combustion chamber,

m gas mass in the combustion chamber,

R gas property constant and

T temperature in the combustion chamber.

1st law of thermodynamic: $\qquad \oint dQ + \oint dW + \oint dU = 0$ [2-3]

with: Q added and released heat energy across the control volume limits,

W mechanical work and

U internal energy.

Balance of masses: $\qquad m_{Cyl} = m_{exh} + m_{inl} + m_{leck} + (m_F)$ [2-4]

with: m_{Cyl} cylinder mass,

m_{exh} mass transferred to and from the exhaust during gas exchange,

m_{inl} mass transferred from the intake during gas exchange,

m_{leck} leackage mass and

m_F fuel mass.

For numerical heat release analysis, equation 2-3 is expressed as differential of crank angle degree φ

$$\frac{dU}{d\varphi} = \frac{dQ_b}{d\varphi} - \frac{dQ_W}{d\varphi} - p\frac{dV}{d\varphi} - A$$

[2-5]

with U: internal energy,

Q_b: heat released,

Q_w: wall heat transfer and

pV: mechanical work.

The last term describes the thermodynamic influence of change of cylinder masses

$$A = -\frac{dm_{inl}}{d\varphi} \cdot h_{inl} + \frac{dm_{exh}}{d\varphi} \cdot h_{exh} - \frac{dm_{vap}}{d\varphi}\left(h_{vap} - h_{evap}\right) + \frac{dm_{leck}}{d\varphi} \cdot h_{leck}$$

[2-6]

with m_{inl}:   intake gas mass,

m_{exh}:   exhaust gas mass,

m_{vap}:   fuel vapor mass,

m_{leck}:   leckage mass, its enthalpies h and

h_{Vap}:   evaporation enthalpy.

For detailed combustion process analysis, the gas exchange (m_{inl} and m_{exh}) is neglectable. The cylinder mass can be determined using test bench measurement data. Especially when looking at turbo charged engines, the zero-dimensional gas exchange calculation gives no more detailed information about the cylinder charge. When looking at direct injecting engines with stratified mixture, the evaporation process has to be modeled thermodynamically (m_{vap}). Otherwise, the 1st sentence of thermodynamic can not be fulfilled [17]. Thus, the vapour enthalpy of the injected fuel is given by

$$h_{vap} = u_{vap} + R_{vap} \cdot T_{vap} .$$

[2-7]

The evaporation process is thermodynamically modeled following Pischinger et al. [18] by

$$\frac{dm_{vap}}{d\varphi}(\varphi_i) = \frac{m_F(\varphi_i) - m_{vap}(\varphi_i)}{\tau} \cdot \frac{dt}{d\varphi}$$

[2-8]

with τ characteristic time of evaporation [s].

2.1.1 The caloric data model approach

The term dU, internal energy, expresses the energetical changes of the cylinder gas depending on temperature, pressure and in-cylinder mixture. Several caloric models have been developed for one typical C/H ratio of fossil fuels [19]. They neglect the dissociation of combustion gases at high temperatures [20],[21] or are only validated for rich mixtures [22]. Grill has shown, that considering the C/H/O ratio of the used fuel leads to improved accuracy regarding the change of internal energy. Therefore, the caloric data of the gas mixture is calculated using an eleven component approach developed by Grill [23]. Under consideration of the equilibrium conditions of eleven gas components, the thermodynamic relevant gas properties are determined. Based on this approach, the specific enthalpy and individual gas constant of the burned gas for arbitrary fuels are determined. Therewith, the fuel composition as well as the temperature dependence of the combustion gas at high temperatures is considered.

Figure 2-1 shows the specific heat at constant volume and the specific internal energy for three C/H/O ratios as a function of temperature. Due to an assumption of a thermodynamical stoichiometrical burned gas despite a global air excess during stratified mixture, caloric data is modeled for a stoichiometrical air-/fuel ratio and a pressure of p = 10 and 50 bar. With decreasing CH-fraction and increasing O-fraction, an increasing specific heat and specific internal energy is found.

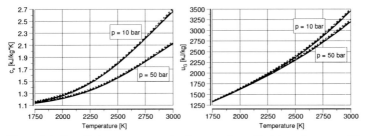

Figure 2-1: Specific heat at constant volume and specific internal energy of the air-/fuel mixture for three C/H/O-Ratios and two pressures as a function of temperature. $\lambda = 1$

Above 1600 K an exponential increase of the specific heat capacity can be observed, dissociation is initiated. With increasing pressure, the effect of dissociation is lowered. Considering engine combustion processes, stratified combustion processes with dethrottled engine operation reach considerably higher combustion pressure compared to homogenous stoichiometrical engine operation. Hence, energy losses due to dissociation are theoretically lower during stratified engine run.

Dissociation mainly proceedes endotherm via the water gas shift reaction [18]

(I) $CO_2 + H_2 \leftrightarrow CO + H_2O$ and   (2-1)

(II) $2H_2O \leftrightarrow H_2 + 2\dot{O}H$,   (2-2)

(III) $2H_2O \leftrightarrow 2H_2 + O_2$,   (2-3)

(IV) $2CO_2 \leftrightarrow 2CO + O_2$ as well as   (2-4)

        $H_2 \leftrightarrow 2\dot{H}$ and   (2-5)

        $O_2 \leftrightarrow 2\dot{O}$.   (2-6)

Molecules are split up into atoms and radicals. The needed energy is detracted from the burned gas, whose temperature decreases. Thermal energy is converted to chemical energy. Figure 2-2 shows the equilibrium constants K for the dissociation reactions (2-1) to (2-4) for dissociation of CO_2 and H_2O as a function of temperature.

According to Pischinger et al. [18], the equlibirum constant K is defined as

$$K_{n,i} = \frac{n_{i,product}}{n_{i,educt}}$$

[2-9]

with   i: reaction index

With increasing temperature the chemical equilibrium is shifted to the product side. Considering the logarithmical scale, an exponential rise of the reaction with increasing temperature and with it an increasing need of thermal energy can be observed. CO_2 dissociates mainly via reaction (2-1) and (2-4) and composes CO despite a stoichiometrical mixture.

Reaction (2-4) shows strong temperature dependence. The back reaction or oxidation freezes at 1600 K under the chosen conditions. Due to chemical losses no complete back reaction can be expected. The energy amount detracted during endotherm dissociation is not completely converted to thermal energy during the oxidation in a combustion engine due to changes of pressure and volume over time.

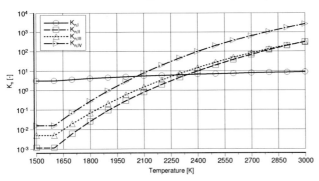

Figure 2-2: Equilibrum constants K for four dissociation reactions (I-IV) as a function of temperature. *p = 50 bar*

Figure 2-3 shows the calculated concentration equilibrium of CO_2, CO, H_2 and H_2O for the chosen condition as a function of temperature for three C/H/O ratios. The CO_2 fraction decreases with decreasing carbon fraction in the fuel and the H_2O fraction increases. The influence of dissociation via reaction (2-2) and (2-3) increases. Assuming constant temperature, via dissociation composed H_2 fraction increases and CO fraction decreases with decreasing C/H-fraction in the fuel.

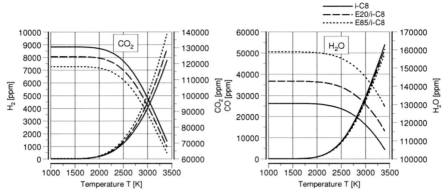

Figure 2-3: Concentrations of CO, CO_2, H_2 and H_2O in the burned gas for three C/H/O-Ratios as a function of temperature. $\lambda = 1$, $p = 50$ bar

Regarding the influence of fuel composition on combustion engine process, a decrease of the C/H-fraction and an increase of the oxygen fraction, like it can be reached via alcohol addition, results in decreased burned gas temperature and therewith lower conversion losses caused by dissociation. The dissociation reaction is shifted to the left side (compare equlibrium constant K (Figure 2-2), less thermal energy is detracted. Additionally, less CO and more H_2 is composed via dissociation due to the reduced carbon and higher water vapor content.

2.1.2 The detailed heat release rate calculation

For heat release calculation, the commercial software TIGER [24] is used. For approximation of the wall heat transfer, the semi-empirical approach developed by Woschni [25] is used according to investigations made by Hoppe et al. [26]. Additionally, the calculation is limited to the high-pressure cycle.

In separation to a working cycle calculation, the heat release is calculated based on the measured cylinder pressure. Due to the use of relative pressure transducers, this measured cylinder pressure has to be corrected for determining the absolute pressure. This is done using the heat release criteria. The released heat Q_b is defined as

$$Q_b = m_F \cdot H_U \cdot \eta_U \qquad\qquad [2\text{-}10]$$

with H_U: lower heat value and

       η_U: conversion efficiency.

Under assumption of no heat release during compression, the 1st sentence of thermodynamic, equation [2-5], is formed to Δp_n

$$\Delta p_n = \frac{\sum_{\varphi=\varphi_{ES}+10}^{\varphi_{soi}-3}\left[\dfrac{dQ_W}{d\varphi} - p_s(\varphi)\dfrac{dV}{d\varphi} + \dfrac{dH_{leck}}{d\varphi} - \dfrac{dU}{d\varphi}\right]}{\sum_{\varphi=\varphi_{ES}+10}^{\varphi_{soi}-3}\dfrac{dV}{d\varphi}} .$$

[2-11]

At first, an approximation value for Δp_n is assumed. Further on, $p = p_s + \Delta p_n$, T, dQ_W, dU and dH_l are calculated leading to a new value for Δp_n. This iteration is done until $\Delta p_n < 1$ mbar and the cylinder pressure can be corrected [27]. Finally, the calculation must be controlled using the energy balance E_B as the ratio of added and released energy in correspondence to equation [2-10] as

$$E_B = \frac{Q_b}{m_F \cdot H_U \cdot \eta_U} .$$

[2-12]

Especially when using split of losses for comparing different engines, this control is unneglectable. Weberbauer [28] has shown that E_B can vary between $95 < E_B(\%) < 101$. Following Woschni et al. [29], this can be reasoned by accuracy deviations of the pressure transducer technology. The tolerances of one measurement setup on one engine should not vary with more than one percent.

2.1.3 The two-zone model for stratified carburetion

For emission reflections using zero-dimensional thermodynamic analysis, a model approach for the temperature of the burned gas is necessary. Therefore, the zero-dimensional control volumina is divided into two zones, the burned gas and the unburned gas. These zones are separated by a thin, massless flame front propagating through the unburned mixture. Because of stratification a heterogenous mixture with many different air-/fuel ratios can be expected which depends on several injector parameters like fuel pressure, injector type and geometry. For example Koch [30] and Hoppe et al. [26] developed a zero-dimensional model approach for fuel stratification and therewith an air-/fuel ratio as a function of time. These models are quite focused on one injector type. For general use during stratification, a homogenous stoichiometrical mixture is assumed around the spark plug. This mixture is surrounded by excessive air and residual gas.

During heat release, the flame front propagates through the unburned mixture until it reaches the surrounding gas. Hence, two border cases can be assumed [17]:

- The burned gas mixes spontaneously with the surrounding air and residual gas likewise to a one-zone-diesel model. The fuel mass flow is considered as already burned as well as proportional to the heat release.

- There is no mixing between the burned gas and the surrounding excess air and residual gas. The burned gas is assumed as stoichiometrical exhaust gas. Additionally, no heat exchange between the burned gas, the excess air and the residual gas takes place. The amount of converted heat is higher than in case one.

Regarding case two, the calculated gas temperature can be separated into a burned and an unburned temperature. The volumina, the mass and its caloric properties as well as the wall heat transfer are calculated for both zones. Thereby, the temperature of both zones can be determined. The burned zone, ergo the burned gas, is assumed to be stoichiometrical. When discussing the resulting burned temperature and emission concentration this simplification has to be kept in mind.

Figure 2-4 shows the specific heat and internal energy for four air-/fuel ratios at constant pressure as a function of temperature. Regarding ignition limits of hydrocarbon fuels the burned gas right behind the flame front mainly varies between $0.8 < \lambda < 1.2$. It can be seen, that below 1700 K specific heats are quite similar in this air-/fuel ratio range. With further increasing temperature the specific heat of a stoichiometrical mixture increases exponentially in the combustion relevant temperature range due to dissociation. The specific heat is considerably higher than under rich or lean conditions (comp. [31]). With leaning burned gas, the initiating temperature of dissociation is shifted to higher temperatures. Enriched burned gas shows a quite different behaviour of the specific heat. Because of the excess fuel enriched burned gas shows a higher specific heat without dissociation. Above 1700 K a decreasing gradient of the specific heat can be observed with a strong increase above 2300 K. Due to oxygen deficit hydrogen and carbon monoxide is composed during combustion and therewith influences the dissociation behaviour of enriched burned gas. Its specific internal energy gradient increases at temperatures above 2700 K. Hence, the simplification of a stoichiometrical burned gas results in a maximum influence of gas dissociation on its caloric properties and therewith the calculation of its temperature.

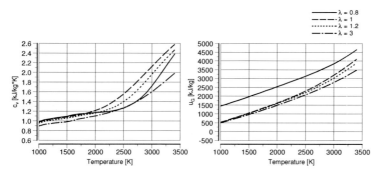

Figure 2-4: Specific heat at constant volume and specific internal energy of burned gas for four air-/fuel ratios as a function of temperature. C_8H_{16}, $p = 50$ bar

Bargende et al. [17] have shown, that using this two-zone model leads to an increased energy balance compared to the one-zone diesel model. The difference in energy balance decreases with increasing residual gas fraction due to changes in caloric properties of the working gas. In the following, this two-zone model is used for estimating burned gas temperature. Consequently, it has to be kept in mind, that the calculated temperature of the burned zone displays the maximum possible gas temperature near to the adiabatic flame temperature due to the assumption of stoichiometrical burned gas [31].

2.2 Combustion of internal combustion engines

Combustion describes the exothermic conversion of an air-/fuel mixture. For complete combustion the reaction equation of a hypothetic fuel $C_xH_yS_qO_z$ can be written as

$$C_xH_yS_qO_z + \left(x + \frac{1}{4}y + q - \frac{1}{2}z\right)O_2 = xCO_2 + \frac{1}{2}yH_2O + qSO_2$$

[2-13]

with the stoichiometrical coefficients

$$x = \frac{M_B}{M_C}c \quad ; \qquad y = \frac{M_B}{M_H}h \quad ; \qquad q = \frac{M_B}{M_S}s \quad ; \qquad z = \frac{M_B}{M_O}o$$

with c, h, s, o – mass fraction of elements contained in fuel.

Considering the mass fraction of oxygen in air, the stoichiometrical air/fuel ratio L_{st} can be written as

$$L_{st} = \frac{1}{0.232}\left(2.664 \cdot c + 7.937 \cdot h + 0.998 \cdot s - o\right)$$

[2-14]

and thus, the air/fuel ratio λ

$$\lambda = \frac{m_{air}}{m_B \cdot L_{st}} = \frac{m_{air}}{m_{L_{st}}}$$

[2-15]

with m_{air} air mass

       m_{Lst} stoichiometrical air mass.

Regarding fuel enrichment ($\lambda < 1$), this is often described as understoichiometric or *rich* combustion. Oxidizer excess ($\lambda > 1$) is described as overstoichiometric or *lean* combustion. Furthermore, the fuel/air equivalence ratio Φ is widely used as reciproce value of λ.

During engine combustion, the molecules C, H_2 and O_2 are ideally converted to CO_2 and H_2O:

$$C_nH_{2n+2} + \frac{1}{2}(3n+1)O_2 \rightarrow nCO_2 + (n+1)H_2O + P$$

(2-7)

Trivially, no ideal complete combustion can be realized. Additionally, next to oxygen, a huge amount of nitrogen can be found in fresh air. Hence, additional products P and radicals are composed during combustion. These are mainly CO, H, OH, O, NO and NO_2. Using equilibrium thermodynamic, the final state of chemical reactions can be described. For detailed analysis of chemical conversion and its influences on combustion, the reaction velocity and the time resolved species concentration is necessary.

A chemical reaction is described with

$$A + B ... \xrightarrow{k_i} E + F + ...$$

with   A,B,...   reactants,

E,F,....   products and

k_i rate coefficient.

Regarding the type of reactions, it can be differentiated between an elementary and a brutto reaction. Meaning an elementary reaction on the molecular level, the reaction proceeds equal to the reaction [32]. During a brutto reaction, intermediate products are composed. The rate coefficient depends directly of the temperature while the reaction proceeds.

This temperature dependence is described by Arrhenius [33] as

$$k = A \cdot e^{\frac{-E_a}{R_m T}}$$

[2-16]

with A: pre-exponential factor,

E_a: energy level of activation,

R_m: gas constant and

T: temperature.

Considering experimental investigations, the equation has been extended by a temperature dependence of the pre-exponential factor

$$k = A \cdot T^b \cdot e^{\frac{-E_a}{R_m T}}$$

[2-17]

with b: temperature coefficient

The energy level of initiation means an energy threshold which has to be exceeded for reaction initiation. During combustion of hydrocarbons, a pressure dependence of the velocity coefficient can appear additionally. Following Glassman [34], this is mostly observed during the decay of long chained radicals into stable species. Combustion processes are based on radical chain mechanisms [32]. As chain reaction is meant to be a reaction, during which a reactive intermediate product - often a free radical - is composed and reacts in several successive reactions further on. The chain reaction can be separated in five types of reactions:

- chain initiation: Active radicals are composed out of stable molecules. For example, this can happen by ionization or thermal excitation [18];
- chain propagation: Active radicals react with other molecules under formation of further radicals. In summation the number of radicals stays constant;
- chain branching: Active radicals react with other molecules under formation of further radicals. During this, there are more radicals formed than consumed;
- inhibition: A radical attacks an already composed product. The product composition is delayed, and
- chain break: Reactive components react to stable molecules.

Regarding the reaction sequence, it is differentiated between an open and a closed reaction sequence [18]. During an open reaction sequence, the formed intermediate product is consumed in the following reaction. This product is not able to interact in a previous reaction. During a closed reaction sequence, a formed intermediate product is able to interfere in a precursing reaction. The reaction always proceeds until the complete conversion of educts. If the reaction velocity increases during a closed reaction sequence, explosions occur. During a thermal explosion, the temperature increases due to exothermic reactions. The reaction velocity increases accordingly and again, more heat is released. Ignition occurs, if the thermal heat release superposes the heat exchange. Considering a homogenous system, this process is described by Semenov´s *Theory of the thermal explosion* [35]. Frank-Kamenetskii has developed the *Thermal theory of explosion* under consideration of spatial inhomogenities of temperature in the reactor system [36]. Newton's law of heat transfer in Semenov's theory is replaced by Fourier's law of heat duction.

During a chain branching explosion, superposition of chain branching reactions against chain braking reactions is the limiting condition for ignition. In opposition to a thermal ignition, during which a sponateneous temperature increase can be observed, an ignition delay occurs during a chemical ignition. This is caused by radical formation during chain branching reactions. The ignition delay time IDT depends on temperature and is mostly described using the Arrhenius approach [18]

$$IDT = A \cdot p^{-n} e^{\frac{B}{T}}.$$
[2-18]

Douaud and Eyzat [37] give an Arrhenius approach for ignition delay time under consideration of the octane number of the used fuel. Simplified, the conversion of hydrocarbons is described by Semenov [38] as

$$RH + O_2 \rightarrow \dot{R} + \dot{H}O_2$$

$$\dot{R} + O_2 \rightarrow Olefin + \dot{H}O_2$$

$$\dot{R} + O_2 \rightarrow \dot{R}O_2$$

$$\dot{R}O_2 + RH \rightarrow ROOH + \dot{R}$$

$$\dot{R}O_2 + R'CHO + R^{\bullet}\dot{O}$$

$$\dot{H}O_2 + RH \rightarrow H_2O_2 + \dot{R}$$

$$ROOH \rightarrow \dot{R}O + \dot{O}H$$

$$R'CHO + O_2 \rightarrow R'\dot{C}O + \dot{H}O_2$$

$$\dot{R}O_2 \rightarrow break - up$$

with   R, R', R'': organic radicals, appear by the abstraction of hydrogen atoms from hydrocarbon compositions and

„•": notes active radicals.

Regarding an ignition by an accelerated chain reaction due to thermal or chemical explosion, it needs to be differentiated between auto and external ignition considering engine combustion. During auto ignition, the needed initiation energy for combustion initiation is realized by compression of the mixture. During external ignition, this energy is added by an external ignition source. The minimum initiation energy is proportional to the necessary ignition volume and its pressure. Consequently, the igniteability of a mixture is limited by its calorical state described via temperature, pressure and specific gas constant.

Also, the type of chemical reactions depends on the caloric state of the mixture. This is universally described by a *p-T-ignition* or *explosion diagram* according to Warnatz et al. [32]. Here, the ignitable areas are separated from the non-ignitable areas using a curve depending on pressure and temperature.

- 1st explosion limit: spontaneous ignition, the production of radicals superposes its stabilization;
- 2nd explosion limit: chain disrupting reactions superpose chain branching reactions, no ignition occurs;
- 3rd explosion limit: thermal ignition threshold, thermal explosion occur.

At lower temperatures and high pressures, the *cold flame* phenomenon occurs. Due to degenerated chain branching reactions, a decelerated combustion proceeds at low temperature. At further increased pressures, multi-stage ignition occurs. This is characterized by several light flashes before ignition. An extended discussion of the explosion limits can be found at Bamford and Tipper [39].

2.2.1 Types of combustion flames

Regarding a classification of technical combustion, it is differentiated between combustion of a premixed and a non-premixed mixture. Furthermore, combustion is characterized by the flow properties of the mixture during combustion. Considering its Reynolds number, combustion is differentiated in laminar and turbulent combustion. Figure 2-5 shows the classification of combustion flames. In a reprocicating internal combustion engine, combustion takes place in a high turbulent regime.

While during idealized combustion investigations the types of combustion flames can be clearly separated, this esperation during engine combustion is much more complex. Regarding the mixture state, under certain operation conditions there is no clear separation between premixed and non-premixed feasible. Therefore, some mixture states are described as partial-premixed combustion with explicit phases of premixed and non-premixed combustion.

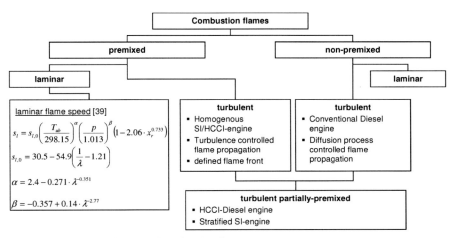

Figure 2-5: Structural diagram of combustion flames.

Turbulent premixed combustion

During premixed combustion, a definite flame front propagates through the premixed air-/fuel mixture. Descriptions of the laminar flame speed, like published by Methghalchi et al. [40], express the influence of pressure, temperature and mixture composition on premixed flame propagation speed. Heat release depends directly of the reaction kinetics and the occurring heat exchange and diffusion processes [41].

Using the Damköhler correlation and flamelet theory, the influence of turbulence and the mean turbulent flame velocity can be expressed. Turbulence bales crumple the laminar flame front. The flame front surface increases. Consequently, more mixture can be consumed and the flame velocity increases. According to Damköhler it is obtained

$$\rho_{ub} \cdot s_t \cdot A_{f,t} = \rho_{ub} \cdot s_l \cdot A_{f,l}$$ [2-19]

with $A_{f,l}$:   entire surface of the crinkled laminar flame front,

$A_{f,t}$:   surface of the mean turbulent flame front and

s_l :   local laminar flame velocity.

The ratio of flame front surfaces is written as

$$\frac{A_{f,l}}{A_{f,t}} = 1 + \frac{v'}{s_l}$$ [2-20]

with v' :   turbulent fluctuation velocity, often described as turbulence intensity.

Hence, turbulent flame velocity can be written as [42]

$$s_t = \left(1 + \left(\frac{v'}{s_l}\right)^{\beta_1}\right) s_l$$

[2-21]

Regarding internal combustion engines, turbulence is nearly proportional to the mean piston velocity, i.e. engine speed [43]. Hence, rate of heat release can be described as

$$\frac{dQ_b}{dt} = \rho_G \cdot A_{f,t} \cdot H_G \cdot s_t$$

[2-22]

with ρ_G mixture density and

H_G mixture heating value.

Due to the turbulence influence on combustion, several characteristic values have been developed for characterizing combustion. A well known approach is a classification using the *Borghi diagram* [42]. For example, Linse et al. [44] match the occurring combustion regimes in a turbo charged SI-engines into the Borghi diagram displaying the influence of engine speed and load on combustion regimes.

Turbulent non-premixed combustion

Turbulent non-premixed flames occur in several technical applications like jet power systems, diesel engines and rocket systems. In a diesel engine, non-premixed flames can be further differentiated considering their ratio of chemical reaction and mixing velocity. If the chemical reaction is fast enough, the combustion process can be defined as mixture controlled. With decreasing pressure and temperature, the chemical reactions are decelerated relative to the simultaneous mixture process. The combustion can be described as kinetically controlled [14]. In opposition to premixed flames, non-premixed flames are characterized by an (ideally unlimited) fast chemistry, which leads quickly to equilibrium. Hence, no defined area, which progresses with a defined velocity, is composed.

Analog to turbulent premixed flames, turbulent non-premixed flames can be treated as local laminar non-premixed flames embedded in a turbulent stream flow. This is described using flamelet concepts. While considering premixed flames, the local laminar and the global turbulent flame velocity can be used for describing the combustion process. Considering turbulent non-premixed flames, the scalar dissipation velocity is used as driving parameter. It describes the increasing deviation from equilibrium with increasing mixing velocity. Under knowledge of the flame structure of laminar non-premixed flames as a function of the scalar dissipation velocity, turbulent non-premixed flames can also be described by using a probability density function (*PDF*) like developed by Libby and Williams [45]. Therewith, the turbulence is described statistically.

Turbulent partial premixed combustion

In several applications, carburetion and reaction occur with similar time scales during combustion. Thus, no clear separation between premixed and non-premixed combustion is possible. These phenomena are described as partial-premixed combustion. An example can be the local flame extinction of non-premixed flames. Fuel and oxidizer have enough time for mixing themselves and burn as premixed flame. [32]

A model approach for partial-premixed flames is the triple flame model, Figure 2-6. Under assumption of a heterogeneous mixture, this is burned with a lean as well as a rich premixed flame and a sharp flame front. Behind this flame front, reduction components like CO from rich combustion as well as oxidizing components from lean combustion zones remain. Assuming satisfying thermodynamic boundary conditions, these components are burned composing non-premixed flames at an air-/fuel equivalence ratio near one [14].

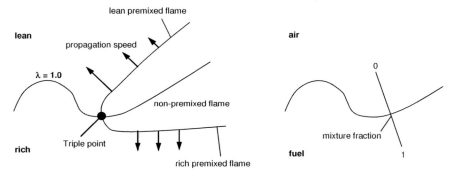

Figure 2-6: Triple flame model according to Merker et al. [14].

2.3 The spray-guided combustion process in direct injection SI engines

The spray-guided combustion process in direct injection SI-engines is characterized by a centrally mounted injector and the possibility of mixture stratification. In differentiation to a wall- or air-wall-guided combustion process, the carburetion of spray-guided combustion processes is mainly characterized by the spray characteristic of the chosen injector [14]. No geometrical advices for guiding the air-/fuel mixture are necessary resulting in improved emission and full-load performance. Until today, two spray-guided combustion processes can be found in production using a piezo-driven injector [4],[5] with a ring-hole nozzle. Several research studies also report about the use of a multi-hole injector [46],[47].

Former investigations have shown the highest efficiency benefit at minimum pollutant emissions when using a piezo-driven ring-hole nozzle [48]. From the outward opening nozzle, the fuel penetrates into the combustion chamber generating a coat shape. Due to the cylinder back pressure, recirculation streams are composed on the outer end of the spray coat enabling a quick air acquisition and mixture formation. For optimum ignition condition, the spark plug is located at this area. Hence, a high degree of spray stability is necessary for stable engine operation. Figure 2-7 shows a scetch of carburetion and resulting air-/fuel ratio structure of spray-guided injection. Due to varying carburetion time scales for individual droplets penetrating into the combustion chamber, a continous air-/fuel ratio gradient is generated [30]. Using multiple injections, the stratification can be modified creating multiple successive mixture rings in the combustion chamber [12].

Figure 2-7: Scetch of carburetion and resulting air-/fuel ratio structure of spray-guided injection.

For illustration, Figure 2-8 shows the CFD calculated fuel mass fraction during a triple injection as an intersection through spark plug and injector. The CFD calculated fuel mass fraction gives information about mixing effects under real engine conditions. The fuel mass fraction is defined as

$$x_F = \frac{1}{\lambda \cdot L_{st} + 1}.$$

[2-23]

All injections create a mixture ring below the spark plug, which is displayed here as two enriched zones in opposition to each other reasoned by the intersectional view. While the first injection evaporizes completely due to the longer remaining time until ignition, the second and main injection results in a still distinctively enriched mixture ring stabilized by the piston bowl. With the third injection stratified upon the second, the spark plug region is enriched and enables the inflammation of the mixture.

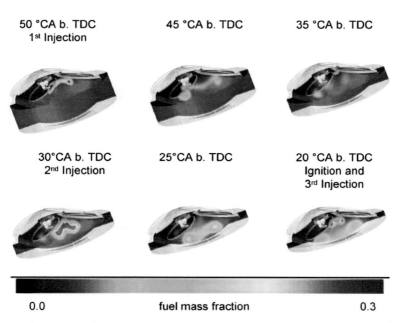

| 50 °CA b. TDC | 45 °CA b. TDC | 35 °CA b. TDC |
| 1st Injection | | |

30°CA b. TDC	25°CA b. TDC	20 °CA b. TDC
2nd Injection		Ignition and
		3rd Injection

0.0   fuel mass fraction   0.3

Figure 2-8: CFD calculated fuel mass fraction during triple injection. n_{mot} = 1500 min^{-1}, imep = 4 bar, λ = 2.8, stratified injection

Comparing Figure 2-8 and Figure 2-9 with strongly enriched mixture conditions, it can be seen that the fuel mass fraction is varied in its intensity due to the higher amount of injected fuel. The local distribution remains quite constant approving a stable mixture formation. A sufficient volume of conditioned mixture is located at the spark plug with a local spark-plug air-/fuel ratio of $0.8 \leq \lambda \leq 1.2$, which enables a stable flame kernel formation. From there on, a primary turbulent premixed flame front propagates through the homogenized mixture areas until the mixture gets too lean or the reaction freezes due to flame and wall quenching effects [31]. Because of a globally high enleaned mixture and areas with not yet carbureted fuel and/or partially burned products like CO, zones with phases of non-premixed combustion occur behind the primary flame front. [49],[50]

Moriyoshi et al. [51] investigated the influence of turbulence on heat release rate. It has been observed, that an increase of turbulence results in increased rates of heat release during the first part of energy conversion approving the dominant effect of turbulence controlled combustion during stratification. Lippert et al. [49] performed combustion visualization on an optical engine with a multi-hole injector. Here, quite good correlations between the CFD model vision and detected combustion phenomena were found. Non-premixed combustion was observed along the spray plumes approving the mixture formation dominated process. Ando et al. [52] and Kuwahara et al. [53] observed a primary blue flame in the ultra violet spectra propagating rapidly through the combustion chamber followed by a second luminous flame with thermal radiation.

| 50 °CA b. TDC | 45 °CA b. TDC | 35 °CA b. TDC |
| 1st Injection | | |

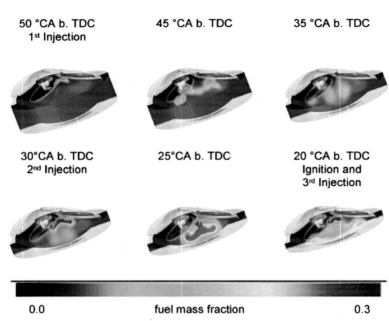

30°CA b. TDC	25°CA b. TDC	20 °CA b. TDC
2nd Injection		Ignition and
		3rd Injection

0.0   fuel mass fraction   0.3

Figure 2-9: CFD calculated fuel mass fraction during triple injection under strongly enriched engine mixture condition. n_{mot} = 1500 min^{-1}, imep = 4 bar, $\lambda < 1$, stratified injection

Several authors [18],[54],[55] showed that, due to the characteristic mixture structure during stratification, partially premixed combustion results in an asymmetrical shape of heat release in comparison to typical premixed combustion of homogenous mixture. Because of slightly enriched mixture around the spark plug, heat release starts very fast reaching high rates of energy conversion. When the flame front reaches the outer areas of the mixture, energy conversion is decelerated distinctively, the conversion of residual products as non-premixed flame can also be characterized by longer timescales. This can result in lower efficiency of the combustion process, if the maintage of heat release takes place before top dead center.

Consequently, during stratified engine operation different emission behaviour can be expected. Figure 2-10 displays the normed HC, CO and NO$_x$ emissions during homogenous and stratified injection as a function of the exhaust gas air-/fuel ratio. As reference point (100%), stoichiometrical, homogenous mixture is chosen. It can be observed, that during stratified engine operation, emissions develop similar like it is known during homogenous engine operation [18]. The range of varation is considerably larger due to the stratified mixture. Respecting further emission relevant engine parameters like residual gas fraction and time position of heat release, the emission curves give information about the mean combustion air-/fuel ratio during stratified engine operation [56]. Remarkable points are the maximum of NO$_x$

emissions or the air-/fuel ratio with increasing CO emissions.

In comparison, it can be noticed that during stratified engine operation NO_x emissions reach a similar level to stoichiometrical, homogenous engine operation with a maximum around $\lambda \approx 2$. Emissions of unburned HC's are considerably lower, also in differentiation to wall-/air-guided combustion processes [57],[58]. Here a minimum at air-/fuel ratio of $1.2 < \lambda < 1.6$ can be noticed. Due to the compact carburetion around the spark plug and the short time between injection and ignition, a high degree of fuel conversion is achieved. Towards leaner operation, HC emissions increase because of an increasing probability of flame and wall quenching [31]. CO emissions are universally considerably lower during stratified engine operation compare to stoichiometrical homogenous engine operation. Due to global oxygen excess, high oxidation rates of CO during stratified engine operation are achieved. At wall-/air-guided combustion processes, also quite higher CO emissions can be noticed [57]. At $\lambda < 1.6$, CO emissions increase comparable to $\lambda < 1.1$ under homogenous mixture condition. The combustion mixture is getting more and more enriched resulting in uncomplete oxidation. Furthermore, at $\lambda < 1.6$ no efficiency improvement can be achieved using stratification [59].

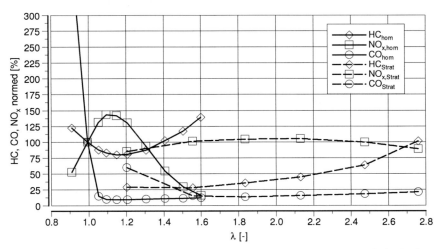

Figure 2-10: Specific HC, CO and NO_x emissions as a function of exhaust gas air-/fuel ratio normed on stoichiometrical, homogenous mixture condition. *1500 min^{-1}, imep = 4 bar, Reference Fuel Ref-O_2-03*

It has to be kept in mind that, during overstoichiometrical engine operation, a standard three-way-catalyst is not able to reduce NO_x emissions. Oxygen excess oxidises too much CO, which is necessary for NO_x reduction. Hence, NO_x emissions are a limiting factor for stratified engine operation. Next to the todays limited pollutant emissions HC, CO and NO_x, possible soot emissions of direct injection SI-engines get into focus of legislation. Due to the given heterogenous mixture structure during stratification, an increased probability of soot emissions can be found during stratified engine operation of a spray-guided combustion process. Since HC and NO_x emissions of direct injection SI-engines are discussed [56],[57],[58],[60] and CO emissions can be directly linked to the air-/fuel ratio, soot formation, oxidation and emission needs to be investigated since soot presents a fraction of particle emissions. The upcoming EU5[+] legislation limits the emitted particle mass and EU6 legislation limits the emitted particle number additionally [61].

2.4 Soot formation, oxidation and emission

During non-premixed combustion soot is formed caused by the diffusion controlled chemical processes and their enhanced timescales. Although there are several reports of soot formation in premixed flames, soot formation takes place in zones of non-premixed mixtures behind the primary flame front (triple flame phenomena). In premixed zones, the needed precursors are consumed before the soot formation process can proceed [32],[62]. By definition, soot are particles, formed in the gas phase of combustion and will be separated from particles originated from deposits in the combustion chamber and further gas bearing engine parts like exhaust gas after treatment systems for example. These are investigated for example by Sandquist [58].

2.4.1 Fundamentals of soot formation and oxidation

Regarding the soot formation process, it is widely accepted that soot evolves from further growing polycyclic aromatic hydrocarbons (PAH), which cluster to soot particles. These consist of more than 10^6 carbon atoms [63] and show quite different shapes and sizes. Thus, they can be described using statistic functions [64],[65]. Chippett and Gray [66] have shown that the statistic distribution can be well approximated using log-normal functions. Smith [67] has reported that "young" soot particle distributions are well approximated via Gauss' error distribution curve and successively migrate to a log-normal distribution due to consecutive agglomeration.

Higher classed hydrocarbons like PAH, which arise after the decomposition in the flame front, have to be generated out of smaller hydrocarbon compounds. The most important precursor is ethyne (C_2H_2), which is mainly composed in enriched zones [68]-[70]. Fuel molecules are cracked thermally. Via the abstraction of hydrogen (H_2) using C_2H_2 and polymerization, polycyclic aromatic hydrocarbons (PAH) are formed. Considering the formation of PAH, Stein et al. [71] described the formation of aromatic ring structures with the reaction of CH or CH_2 with C_2H_2 under formation of C_3H_3, which can form the first ring using recombination and relocation [72]. Frenklach and Wang [65] developed an elementary reaction mechanism, in which several molecules of the ethyne generating rich combustion merge to a first benzol-circle. In doing so, hydrogen is abstracted and single hydrogen atoms are accumulated. Under proceeding hydrogen abstraction and ethyne accumulation (HACA mechanism [73]) coherent PAH circles are formatted, which agglomerate (Agglomeration: Surface growth by adhesion of smaller particles) to larger spherical entities [68],[74],[75]. Figure 2-11 shows the schematic reaction path of soot formation described by Bockhorn [75].

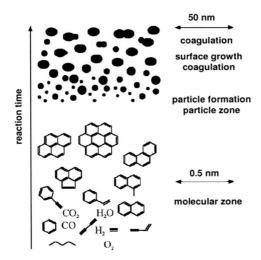

Figure 2-11: Schematic reaction path of soot formation in homogenous mixtures or premixed flames. Taken from Bockhorn [75].

Via surface growth by addition of mainly C_2H_2 and coagulation (Coagulation: Collision and fusion of particles), three dimensional soot particles are formed. The particle inception takes place near the flame front [76]. Finally, the coagulation superposes the surface growth and leads to deceleration of the particles' growth due to the reducing particle surface. Depending on the gas temperature, the surface growth is disrupted [76] after consumption of the available hydrocarbons. Additionally, soot particles can be formed by coagulation of soot nuclei without surface growth and build spherical primary particles [77]. After coagulation, the particles cluster further on and branching soot agglomerates are generated. The degree of clustering predominantly defines the optical properties of soot particles, which influences the optical analysis of soot particles [66],[78]-[80]. Furutani et al. [81] describe the dependence of soot particle size on the carbon/hydrogen (C/H) ratio. The particle diameter increases linearly with the C/H ratio during soot formation. The authors conclude that hydrogen can only be found on the surface of soot particles. Richter et al. [82] give an overview considering several studies of PAH's growth to soot.

Considering thermodynamically combustion of hydrocarbons in fuel enriched mixtures, soot formation would only be able to occur at carbon/oxygen (C/O) ratios higher than one. In opposition to that, Haynes and Wagner [74] have shown for hydrocarbon/air mixtures that soot formation occurs at much lower C/O ratios. Thus, soot formation is kinetical controlled. Haynes and Wagner also classified hydrocarbons regarding their tendency towards soot formation [83]. Figure 2-12 shows a classification of hydrocarbons regarding their tendency towards soot formation. Accordingly, soot formation tendency increases proportionally to fuel molecule size. Likewise the degree of branching or multiple bondings support soot formation in opposition to stretched molecules. Aromatic bondings show the highest soot formation tendency, they include the soot precursor of PAH's.

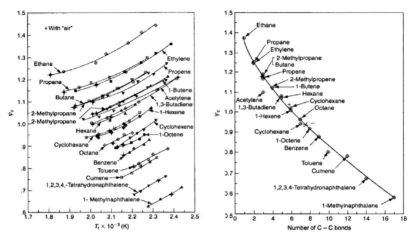

Figure 2-12: Classification of hydrocarbons regarding their tendency towards soot formation. Taken from Glassmann [34].

The rate of soot formation is mostly described by the soot volumina fraction. It has been shown by several authors [32],[75], that the soot volumina fraction increases with increasing pressure and C/O ratio while the temperature dependence can be described using a normal probability curve. For soot formation, radical precursors are needed. Thus, soot formation is limited to lower temperatures. Towards higher temperature, the oxidation rate rapidly increases. A maximum of soot formation rate can be found between 1600 and 1700 K for premixed C_2H_4 flat flames [84].

Figure 2-13 shows a Φ-T-map with the area of soot and nitric oxides formation calculated for combustion engine boundary conditions [85]. Based on kinetical considerations [31],[75], soot formation mainly occurs with local air-/fuel ratios lower than $\lambda = 0.34$ (according to a mixture C/O ratio higher than one). Already Street et al. [86] and later for example Hansen [87] report that even at $\lambda = 0.6$ first soot formation occurs. The lower temperature limit for soot formation is described at T = 1500 K. Under engine condition with increased pressure and temperature, an increased soot formation towards lower air-/fuel ratios can be found. The particle number increases while the particle size remains constant [88] compared to atmospherical conditions. Tree et al. [89] assume that the influence of coagulation decreases. The velocity of coagulation decreases, because the length scale of the gas becomes less than the the particle diameter. Hence, the molecular motion decreases. Regarding pressure influences, soot formation increases linearly with the pressure.

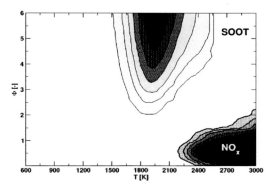

Figure 2-13: Φ-T-Map for soot an nitric oxides formation during hydrocarbon/oxygen combustion. Taken from Kook [85].

Contemporaneous to soot formation, soot particles are oxidized under formation of CO and CO_2 if sufficient oxygen and temperature is present [82],[90],[91]. Due to its aromatic basic structure soot particles are quite stable. After oxidization of smaller hydrocarbons, soot particles are oxidized starting with highly volatile PAH's [92]. Since oxidation is an exothermic process, the surface temperature of oxidizing particles is expected to increase. Oxidation drivers are oxygen, carbon mono- and dioxide, water vapour and nitric oxide. Nagle and Strickland-Constable [93] have developed a semi-empirical correlation for the oxidation of soot particles by oxygen, which has been proved by Park and Appleton [94]. At constant oxygen partial pressure, the oxidation rate exponentially increases with the temperature. During combustion, at suitable sufficient temperatures the composed OH radicals and oxygen atoms consume formed soot particles. Under engine combustion condition, the hydroxyl radical is identified as a dominant oxidizer [34],[82],[95]-[97]. Puri et al. [97] have shown that soot successfully competes with CO for OH radicals and therewith, soot oxidation suppresses CO oxidation despite CO's higher reactivity and neglecting additional CO formation. A further overview on models of soot formation and oxidation can be found at Kennedy [91].

2.4.2 Soot formation, oxidation and emission in internal combustion engines

Because of non-premixed combustion as the dominant type of combustion, soot formation during engine combustion is widely investigated in diesel engines [87],[98]-[106]. In direct injection spark ignition engines, soot is formed due to unsatisfying caburetion [14],[18],[88]. During homogenous injection mode, this is mostly caused by interaction of fuel spray with combustion chamber walls and/or inlet valves and the oscillating piston [107]. Cold start engine conditions increase the probability of fuel wetting and subsequent pool fires [108]. The need of fuel enrichment at high engine loads and speeds increases the probability of soot formation due to a local lack of oxygen also in the gaseous phase [31]. The injected and interacted fuel builds up carbon deposits which are burned under extremely rich and non-premixed condition, soot particles are formed [107]. Pischinger et al. [99] split soot formation and oxidation during diesel engine combustion in four periods:

- At first soot is formed in extremely fuel enriched mixture zones due to the short time for carburetion after injection.
- In a second period soot is formed during injection in the already propagating flame.
- A third way of soot formation is caused by fuel injection in the burnt gas. At temperatures about T=1500 K, soot formation can be observed.
- Finally, formed soot particles are oxidized in the combustion chamber.

Soot oxidation during engine combustion occurs simultaneously and competes with soot formation. According to Kittelson et al. [109], about ninety percent of formed soot during combustion is oxidized. In diesel engines, soot oxidation could be observed at temperatures above T=1800 K [110]. Analog to theoretical considerations in [93], the rate of oxidization increases exponentially with temperature. Several authors have reported that the oxygen content of fuel correlates linear with soot emission due to an improved oxidization [98],[103],[105],[111]-[116]. It is reported that gasoline fuels lead to smaller particles and less agglomerates compared to diesel fuels. Franke and Tschöke [117] have proved that the mean diameter of the formed soot particles depends on fuel composition.

With the use of mixture stratification under overstoichiometrical conditions, the rate of non-premixed combustion in a gasoline DISI-engine and with it the occurrence of soot formation during combustion is widely inevitable due to local enriched zones around the spark plug. Regarding inner engine soot formation mechanisms following Pischinger et al. [78], soot is formed mainly during non-premixed combustion in extremely rich zones and by flame interaction of liquid fuel droplets during stratified mixture operation. Figure 2-14 shows an overview of possible sources for soot formation.

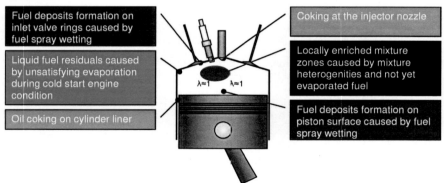

Figure 2-14: Sources of soot formation in direct injection SI-engines.

Summarizing, soot is composed under local highly enriched mixture conditions. Its formation takes always place under non-premixed conditions due to its high timescales, even if the combustion process is dominated by premixed flame propagation. Important to notice, that soot oxidation reduces up to ninety percent of the composed soot and is mainly controlled by the occurrence and consumption of hydroxyl radicals. Thus, smoke emissions depend on soot composing conditions as well as on the ability of their oxidation. In a spray-guided combustion process, zones and phases of non-premixed combustion with its characteristic formation of soot particles are nearly unevitable. For avoiding the emission of pollutant smoke emissions, the combustion processes and the corresponding emission formation and oxidation need to be controlled and understood in detail. Due to the kinetical character of soot formation, flame temperatures of non-premixed combustion are significant for controlling its formation. Even more, possibilities of oxidation enhancement like achieved by ethanol enriched fuels need to be discussed.

2.5 The optical determination of non-premixed flame temperature

For an improved investigation of physical and chemical processes in internal combustion engines, optical analysis is unavoidable. Already in the 1930 first reports on combustion visualization have been published. Glyde et al. [118] published first records of the combustion on an L-head SI-engine. Withrow and Glassweiler [119] implemented successfully a quartz window into a similar engine in 1936. Using a high-speed camera they were able to record thirty images of a working cycle at a frame rate of 500 fps and to give first explanations of knock phenomena. Today, several qualitative and quantitative measurement technologies are established. For combustion visualization and analysis, radiant emission of chemiluminiscence or luminosity due to chemical reactions or body radiant emission is used mostly. Next to the visualization, combustion imaging is used for determining combustion relevant data and species like the air-/fuel ratio, emissions concentrations and gas temperature. Radiant emissions are classified by Gaydon [120] in

- line spectra,
- band spectra and
- continuous spectra.

Using line spectra, radiant emissions with discrete wavelength are described. They occure during the transience of single atoms between discrete energy levels. Band spectra are diffused enlarged and continuous molecular spectra, which occur during energy level transience in molecules. One example is the chemiluminiscence of radicals in the reaction zone of a flame. Continuous spectra are emitted by solid or fluid bodies. It is characterized by all wavelengths at different intensities.

Fundamentally, all types of radiant emission exist during combustion [60]. Line spectra are less important. During premixed combustion, only chemiluminiscence of band spectra appears. During non-premixed combustion, solid bodies (mainly soot) are generated. These emit continuous spectra. Their intensity is much higher than the chemiluminiscence of premixed combustion.

During exothermic heat conversion of molecules, electronic exaggerated atoms or molecules are generated, which are not in thermal equilibrium. Chemiluminiscence appears, if the exaggerated molecules fall back spontaneously into their ground state. Thus, they emit energy in form of radiant emissions. Those are described by Plancks' quantum of action and the frequency of the radiant emission [121]. The product of Plancks' quantum of action and the radiant frequency has to be equal to the difference of the exaggerated and the basic energy level (Bohr´ frequency condition). According to Schänzlin [54], collision quenching of molecules can occur.

This leads to transience between the energy levels without radiant emission. Thus, chemiluminiscence only occurs during spontaneous decayed photons. Reffering to Docquier et al. [122], Ohmstede [123] and Abata et al. [124], OH, HCO, C_2, and CH_2O are the noticeable radiating components in the visible and ultra violet spectra, CO_2, CO, und H_2O the noticeable components in the infra red spectra during premixed combustion. Aust et al. [125] used an emission-absorption spectroscopy for crank angle resolved temperature determination in a port-injected SI engine. Further investigations have been published by Reissing [60], Achleitner [126] and Gaydon [120].

Body radiant emission is characterized by a continuous spectrum. Ideally, body radiant emission is defined as a black body, a so called black radiator. According to Kirchhoff, the spectral degree of emission ε (λ, T) of a black radiator is equal to the spectral degree of absorption α (λ, T) at given wavelength and temperature. Using Planck´s law of radiant emission, the sprectral radiant density $I_{\lambda,S}$ of a black radiator, common to the expected distribution of intensity, can be described as [127]

$$I_{\lambda,s}(\lambda,T) = \frac{c_1}{\pi \cdot \lambda^5} \cdot \frac{1}{e^{\frac{c_2}{\lambda T}} - 1}$$

[2-24]

with c_1 $= 3.74 * 10^{-16}$ $[W^*m^2]$ (first Planck constant)   and

   c_2 $= 1.44 * 10^{-2}$ $[m^*K]$ (second Planck constant).

The wavelength with maximum intensity depends on the temperature. This is described by Wien's law [128]

$$\lambda_{max} \cdot T = 0.2898 \qquad [cm^*K].$$

The whole emission S is given by integration of the radiant emission power. This is decribed by the Stefan-Boltzmann law [129],[130] depending on the temperature

$$S = \sigma \cdot T^4$$

with $\sigma = 5.67 * 10^{-8}$ $[W/m^2 {}^*K^4]$.

The black radiator has to be seen as a theoretical and idealized case. In practical applications, coloured or grey radiators occur. The deviation of these radiators concerning their emissivity is decribed by the degree of emission ε (λ).

This is defined as

$$\varepsilon(\lambda) = \frac{I_\lambda}{I_{\lambda,s}},$$

[2-25]

the quotient of the radiance emission density of a coloured or grey radiator and a black radiator. Consequently, ε (λ, T) is one for the black radiator and ε (T) is minor one for a grey radiator. Thus, it is assumed, that the degree of emission is constant concerning to the wavelength. When considering a coloured radiator, the degree of emission is also minor one, but it also depends on the wavelength.

2.5.1 Optical determination of non-premixed flame temperature using the two colour method

For optical determination of temperature, several measurement technologies are known. A superior classification can be made either the chemiluminiscence of chemical reactions during combustion or the quantum emission of particles is detected and whether chemiluminiscence is inducted using light or heating sources like pulsed lasers. Zhao and Ladommatos give a great survey of common technologies and corresponding publications [131].

Using laser light-sections, gas temperature can be determined due to the excitement of gas species. Their decay behaviour after exaggeration gives information about their temperature under assumption of thermal equilibirum. Using LIF, Kuhn determined flame temperature profiles of methane flames [132]. Furthermore, gas temperature can be determined by the addition of elements like defined metals (for example phosphor or indium), which are exaggerated by laser light. Measuring the decay time, the temperature can be determined. Using laser light-sections, also two-dimensional distributions can be imaged [134]. For determining the flame temperature of premixed combustion, a further common approach is the *Raman-Spectroscopy*. Using a pulsed laser source, gas molecules are excited and emit light energy, which can be detected. A further advanced approach is the *Coherent-Anti-Stokes-Raman-Spectroscopy (CARS)* (for example Brackmann et al. [133]). Reissing et al. [135] determined the gas temperature of premixed combustion using a spectroscopical detection of the alkali Sodium and Potassium. Aust et al. [138] used a laser intensified spectroscopy of Potassium for determining the gas temperature. On the other hand, Sakurachi et al. [136] used an infra red pyrometer with a narrow band pass filter for CO_2 to determine the gas temperature of premixed combustion. Li and Gupta [139] demonstrated the application of a photo thermic deflection spectroscopy for premixed gas temperature identification using a gas burner.

Non-premixed combustion phases are characterized by the formation of soot particles, which emit continous spectra. Due to its dominant occurrence during Diesel combustion, optical measurement technologies considering soot concentration and temperature are widely used on Diesel engines. A survey can be found at Zhao and Ladommatos [140]. Using laser technologies, laser induced incandescence (LII) and light scattering are widely used for determining soot concentrations and their temperature. For example, Geitlinger et al. [141] used a combination of LII and Rayleigh scattering for the determination of soot volume fraction, particle number densities and their size in laminar and turbulent burner flames. Dec et al. [142] applied a combination of LII and laser induced scattering (LIS) on an optical single-cylinder DI Diesel engine.

The soot pyrometry using the two colour method is a passive, non-invasive measurement technology for determining the non-premixed flame temperature. Depending on the calculation method and its assumptions, even the soot concentration and soot distribution during combustion can be determined. The application is limited on non-premixed flames [32]. According to chapter 2.4, during non-premixed combustion carbon particles are formed in the flame. These are heated up due to a high convective heat transfer between the particles and the surrounding gas. Because of this, they emit light with a continous spectrum. Based on Planck's law, the temperature of these glewing particles can be calculated. The emitted light is filtered at two different wavelengths (two colours) and detected. Calculating the ratio of the intensity of the emitted light, it can be linked to a temperature [131]. Due to the high convective heat transfer, it can be assumed that flame temperature is equal to particle temperature. Already in the 1940ies, this characteristic was used for the determination of flame temperature [131]. Since the early 1980ies, the two-colour method is a widely used application for determining flame temperature in diesel engines [78],[80] [98],[100],[101],[106],[110],[143]-[156].

Referring to gasoline SI engines, using stratified air/fuel mixture for part load condition as well as homogenous mixture under high pressure turbo charging full load condition, zones of non-premixed combustion can be detected caused by unsatisfying carburetion or partial-premixed mixture. Hence, it is possible to visualize these combustion phases and estimate the flame temperature using the 2-colour method. Most investigations performed used optical fibre probes to gain zero-dimensional information about soot concentration and temperature. Mayer [78] and Reissing et al. [135] used an optical fibre probe on a single-cylinder SI engine for determining soot concentration and temperature in comparison to a DI Diesel engine. Non-premixed Diesel combustion resulted in quite higher luminosity intensity compared to stratified SI engine combustion. Schänzlin [54] applied several optical measurement applications on a single cylinder optical engine with a 1st generation spray guided combustion process. Using the extended two-colour method with three optical fibres, influences of engine parameters like rail pressure and injection time on non-premixed combustion and the its corresponding soot concentration and temperature are displayed. Wyszynski et al. [157] used a single-cylinder Mitsubishi GDi engine with a quartz piston to get a maximum view on combustion processes. Using a three-colour method, soot concentration and temperature were imaged two-dimensionally. Stojkovic [158] developed a high-speed imaging system on a single-cylinder DISI engine with quartz piston and wall-guided combustion process. Using the two-color method, the formation of soot during stratified engine operation can be described. Additionally, hydroxyl radicals are detected for describing soot oxidation effects. Ma et al. used a CCD camera system [159] as well as a high-speed camera system [160] for imaging soot luminosity and determining soot concentration and temperature on a single-cylinder DISI engine with quartz piston and centrally mounted injector.

Nevertheless, the two-colour method implicates some limitations. The measured light intensity is integrated over the optical beam. Hence, two-dimensional images of the spatial flame structure are projected. Naccarato [161] developed a numerical reconstruction of the third dimension using a neural network trained by images of the piston position to solve this limitation. Furthermore, flame temperature estimation is limited to a certain temperature range. Towards higher flame temperature, soot particles are too small to emit detectable light and are pyrolized. Towards lower flame temperature, colder particles with a low light emission in the near infrared spectra cannot be detected. Especially when calculating the soot concentration this leads to an unneglectable error. Next to the influences of the optical setup, the temperature calculation can be differented in two methods, the relative method and the absolute method [162].

The relative method

The relative method describes the light emission ratio assuming soot as a grey radiator. Hence, the degree of light emission ε does not depend of the considered wavelength. Therewith, an explicit temperature determination is possible [162]. The radiation intensity is described using Wien's approximation [128] of Planck's law of radiation [140]

$$I_\lambda(\lambda,T) = \varepsilon(\lambda) \cdot I_{\lambda,s}(\lambda,T) = \varepsilon(\lambda)\frac{c_1}{\lambda^5 \cdot e^{\frac{c_2}{\lambda T}}} \quad .$$

[2-26]

With the assumption of a grey radiator the ratio of two radiation intensities can then be written as

$$\frac{I_{\lambda_1}(\lambda,T)}{I_{\lambda_2}(\lambda,T)} = \frac{\lambda_2^5\left(e^{\frac{c_2}{\lambda_2 T}}\right)}{\lambda_1^5\left(e^{\frac{c_2}{\lambda_1 T}}\right)} \quad .$$

[2-27]

So the ratio can be linked directly to a temperature as follows

$$T_{rel} = \frac{c_2\left(\dfrac{1}{\lambda_2} - \dfrac{1}{\lambda_1}\right)}{\ln\left(\left(\dfrac{\lambda_1^5}{\lambda_2^5}\right)\dfrac{I_{\lambda_1}(\lambda,T)}{I_{\lambda_2}(\lambda,T)}\right)} \quad .$$

[2-28]

The absolute method

The absolute method describes an implicit determination of temperature [162] assuming a colored radiator with a wavelength depending degree of emission [163]. Therefore, a virtual temperature, the so called apparent temperature T_a, is introduced. The apparent temperature T_a is assumed as a temperature of a black body which radiates the same intensity like the colored radiator with the temperature T

$$I_{\lambda,s}(T_a) = I_\lambda(T) \quad .$$

[2-29]

Through conversion of Planck's law of radiation the spectral degree of emission can be written as

$$\varepsilon_\lambda(\lambda,T,T_a) = \frac{e^{\frac{c_2}{\lambda T}} - 1}{e^{\frac{c_2}{\lambda T_a}} - 1} \quad .$$

[2-30]

Analog to Beer-Lambert's law, Hottel and Broughton [164] describe the spectral degree of emission with an empirical correlation

$$\varepsilon_\lambda(\lambda) = 1 - e^{\left(\frac{-KL}{\lambda^\alpha}\right)}$$

[2-31]

with K absorption coefficient proportional to the numbers of soot particles per volume segment and

L   the geometrical thickness of the flame along the optical axis.

The parameter α depends on the physical and optical properties of the soot in the flame. Via equalization of both equations the KL factor can be described with

$$KL = -\lambda^{\alpha} \ln\left[1 - \left(\frac{e^{\left(\frac{c_2}{\lambda T}\right)} - 1}{e^{\left(\frac{c_2}{\lambda T_a}\right)} - 1}\right)\right]$$

[2-32]

Similar to the relative method, this equation can be formulated for two different wavelengths and be set to ratio. Therewith, the unknown KL factor is eliminated and the equation can be reduced on a zero-solving equation [98]

$$\frac{\lambda_1^{\alpha_1}}{\lambda_2^{\alpha_2}} - \frac{\ln\left(1 - \frac{e^{\frac{c_2}{\lambda_1 T}} - 1}{e^{\frac{c_2}{\lambda_1 T_{a,1}}} - 1}\right)}{\ln\left(1 - \frac{e^{\frac{c_2}{\lambda_2 T}} - 1}{e^{\frac{c_2}{\lambda_2 T_{a,2}}} - 1}\right)} = 0$$

[2-33]

Schubiger [98] suggests a numerical solver using the reduced Newton-Raphson method. The two apparent temperatures $T_{a,1}$ und $T_{a,2}$ need to be calibrated.

As described the choice of the empirical parameter α is quite important for temperature calculation using the absolute method and the estimated soot concentration. According to Zhao und Ladommatos [131], α is a function of the wavelength, the soot particle size and the refractive coefficient of the soot

$$\alpha = f(\lambda, d_{Soot}, m)$$.

Gaydon and Wolfhard [165] also found a dependency on fuel properties. Regarding the sensitivity of α, Zhao and Ladommatos describe a strong dependency on the calculated temperature when using wavelengths in the infrared spectra. When using filters with wavelengths in the visible spectra only small sensivities can be observed. Regarding the sensitivy of the calculated KL factor α, an oppositional behaviour can be observed. A review on used values for α in literature can be found at Zhao and Ladommatos [131].

2.5.2 The 2-ColourRatioPyrometry (2-CRP)

All to the author known investigations of non-premixed combustion in gasoline DISI engines using the two-color methods til today are limited to single-cylinder engines with quartz pistons or the use of zero-dimensional optical fibre probes. For experimental combustion process research, the efficient usability of optical measurement technology and its reliability are the most important factors. Therefore, an optical measurement system needs to be stable and applicable on entire research engines without fundamental modifications. During these investigations, the 2-ColourRatioPyrometry (2-CRP) is introduced. Next to its optical setup it can be characterized by high usage stability and an easy integration into test bench measurements.

Experimental setup of the 2-ColourRatioPyrometry

The built system consists of

- a double image optic concluding a filter unit,
- an intensifying relay optic (IRO),
- a CCD chip camera and
- an user PC.

The object to be imaged is focused via a lens (L1) to the image plane. There, a rectangular aperture limits the image area. The second lens (L2), arranged behind the aperture, generates a parallel light, which is seperated into two images inside the fourier-plane using a prism. The camera system is mounted exactly along the optical axis. The two images outside the fourier-plane are mapped via a third lens (L3) to the right and left side of the image plane. Therewith, a virtual twin camera system is developed with two images filtered at different wavelengths.

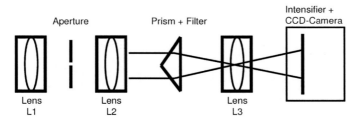

Figure 2-15: Double imaging optic of the 2-CRP system

The camera system consists of an intensified relay optic (IRO) and a standard CCD camera. The IRO intensifies the two images and sends them onto the CCD camera chip. Technical details can be found in appendix k. On the camera chip, the two images are separated using a software algorithm and recorded as two images on the user PC. The filter unit is placed between the prism and objective 3 (O3).

To match the two images and calculate the signal ratio R, the optical setup needs to be calibrated perspectively. Therefore, an image of a plane is taken with round dots plotted on it (compare Figure 2-16). Using a software wizard, a triangle of three dots on each camera view is marked manually.

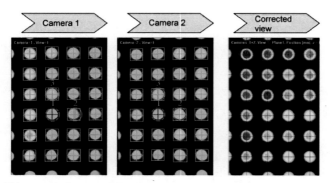

Figure 2-16: Camera views of perspective calibration using a calibration plate

Based on this information, a software algorithm identifies the rest of the dots on each camera view and matches the view. Here the quality of the calibration image and an exact marking of the imaged dots are quite important. The more dots can be identified on both camera views, the better is the match. Hence, the perspective distortion can be calculated and corrected using a pinhole model calibration [166].

Image data processing of the 2-ColourRatioPyrometry

Prior to the temperature calculation itself, a detailed post processing of the recorded raw image is necessary. Because of the use of a CCD camera, the image can be described as an arrangement of intensities written in a matrix analog to the number of pixels available on the camera chip. From there on, the post processing consists of several operations applied on this light intensity matrix (1). Figure 2-17 shows the flow graph of the image post processing. All shown operations are performed using the software DaVis 7.1.

At first, the images are corrected by executing background subtraction and chip inhomogenity suppression. Therefore, a background image and a sheet image are recorded without the filter unit on an optical test bed. For background substraction an image with ambient light is recorded and its intensities substracted from the measurement record. For chip inhomogenity suppression a sheet image is taken of a homogenous light source and the images are divided by each other. Using this ratio, inhomogenities on the image intensifier and the CCD chip are suppressed. Then the splitted two images are fit to each other using the described perspective correction. For reducing calculation costs, a mask application is executed for excluding all pixels, which do not show the image transported by the endoscope lens. Then, finally, the RB-ratio is calculated and therewith the temperature can be determined (3).

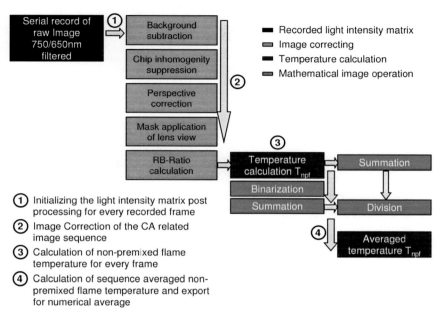

Figure 2-17: Image post processing of 2-CRP.

Because of the limited repetition rate of the CCD camera, only one image per cycle can be recorded. To suppress the influence of cycle-to-cycle variations, several images at every crank angle are recorded. Hence, an averaging of this single images need to be processed. Therefore, the temperature images are once binarized and summarized. The same images are additionally summarized and then divided by the binarized and summarized image. By this, the average for every pixel can be determined (4). For further numerical evaluation, the over-image average needs to be calculated. Here, a MATLAB code is applied calculating a weighted average temperature of the image. For the average value, only the pixels are averaged which show a temperature value. By this, the pixels, showing no non-premixed combustion, are excluded.

Temperature calculation approach of 2-CRP

The 2-ColourRatioPyrometry used in this work performs the relative method for calculating the non-premixed flame temperature. The temperature calculation is implemented in a post processing macro using DaVis 7.1 provided by LaVision GmbH [167]. According to this, the light detected by the camera system can be described as the product of the transmission curve of the band pass filter $F(\lambda)$, the sensitivity of the camera $Z(\lambda)$ and Wien's formula $I_\lambda(\lambda,T)$

$$S(\lambda,T) = \int F(\lambda) \cdot Z(\lambda) \cdot I_\lambda(\lambda,T) d\lambda .$$

$$[2\text{-}34]$$

Figure 2-18 shows the theoretical signal $S(\lambda,T)$ for the used filter wavelength (λ_1=750nm, λ_2=650nm) as a function of the temperature.

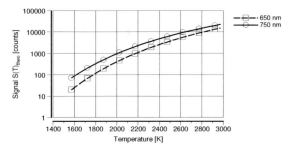

Figure 2-18: The signal S as a function of temperature for the used filter wavelengths.

For narrow band filters like used here, the integral can be approximated to

$$S(\lambda,T) \approx \Delta\lambda \cdot F(\lambda) \cdot Z(\lambda) \cdot I_\lambda(\lambda,T).$$

[2-35]

When imaging the radiation at two colours using two filters with different wavelengths λ_1 and λ_2, the two signals can be described as the ratio $R(\lambda_{1,2}, T)$

$$R(\lambda_{1,2},T) = \frac{S_m(\lambda_1)}{S_m(\lambda_2)} = \frac{\Delta\lambda_1 \cdot F(\lambda_1) \cdot Z(\lambda_1) \cdot I_\lambda(T,\lambda_1)}{\Delta\lambda_2 \cdot F(\lambda_2) \cdot Z(\lambda_2) \cdot I_\lambda(T,\lambda_2)}$$

[2-36]

with the product of $\Delta\lambda*F(\lambda)*Z(\lambda)$ depending only on the filter used and the camera detector response. Hence, this can be described as a constant factor in the measured ratio C_m

$$C_m(\lambda_{1,2}) = \frac{\Delta\lambda_1 \cdot F(\lambda_1) \cdot Z(\lambda_1)}{\Delta\lambda_2 \cdot F(\lambda_2) \cdot Z(\lambda_2)}.$$

[2-37]

Describing Wien's formula with

$$c_{12} = \left(\frac{\lambda_1}{\lambda_2}\right)^{-5} \qquad \text{and}$$

[2-38]

$$A_{12} = c_2 \cdot \frac{\lambda_1 - \lambda_2}{\lambda_1 \cdot \lambda_2},$$

[2-39]

the ratio $R(\lambda_{12},T)$ can be written as

$$R(\lambda_{12},T) = C_m c_{12} \cdot e^{\left(\frac{A_{12}}{T}\right)}.$$

[2-40]

Hence, the temperature T_{npf} can be calculated via

$$T_{npf} = \frac{A_{12}}{\ln(C_m c_{12}) - \ln(R_{12})}.$$

[2-41]

For solving this equation, the constants A_{12} and $C_m c_{12}$ need to be calibrated. This is done using a tungsten ribbon lamp, for which the temperature of the tungsten ribbon is known and related to a certain electric current. The ratio R is measured at different tungsten ribbon temperatures and the calibration constants are determined using a linear regression like shown in Figure 2-19. Following equation 2-40, the relation of ratio R and temperature T can be written as

$$\ln(R(\lambda_{12}, T)) = \ln(C_m c_{12}) + A_{12} / T_{npf}$$
[2-42]

Using the linear regression, the fit values A_f and $\ln(B_f)$ can be related directly to the calibration constants

$$A_f := A_{12} \quad \text{and}$$

$$\ln(B_f) := \ln(C_m c_{12}).$$

Figure 2-19 shows equation 2-42 as a function of the inverted temperature. The constants are taken from technical specifications of the optical system. Thus, a line results with a determination coefficient of $R^2 = 1$ like expected for the model fit.

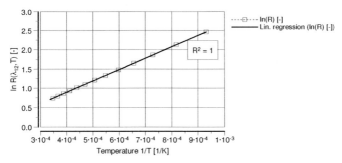

Figure 2-19: Example of the logarithm of calculated ratios using theoretical specifications versus the inverted temperature and the linear regression and their determination coefficient R^2.

Temperature calibration for engine measurement

For on-engine temperature determination, reference calibration is necessary. Using a tungsten ribbon lamp, the intensity ratio at different temperatures is determined giving the signal ratio R as a function of temperature of a reference body. Based on this, the needed fit values for the measured temperature range are determined and are used for temperature calculation form the taken records.

Because of optical limits, the available sensitivity of the optical setup is limited to a certain temperature range under laboratory conditions (ambient temperature and pressure, room conditions).Thus, only a temperature range of 500K can be calibrated with one parameter setup. The parameter setup mainly contains of gate and gain settings. Varying gain during calibration yields to unacceptable calibration tampering. Due to this, several temperature ranges are calibrated at constant gate and gain. To avoid optical saturation or too high signal noise, the gate of the IRO is adjusted. The gain is kept constant at seventy percent for all measurement operations. Figure 2-20 shows the calculated signal ratio for three calibration images recorded at three different gates. All three calibration curves show quite similar developing versus changes of temperature of the tungsten ribbon lamp. At the high and low temperature end of the recorded temperature range, strong gradients occur resulting in some inaccuracies. Especially at low temperatures, several different signal ratios are found for one temperature. Hence, no clear temperature determination is possible. These interferences are reasoned by saturation effects at higher temperatures and strong noise effects at low temperatures. Although background and inhomogenity suppressions are performed before calibration, the constant factor C_m cannot be assumed as constant under real measurement conditions.

To gain a trustworth evaluation, a fitting function is necessary. By this, effects of signal noise and geometrical inaccuracies of the perspective calibration on ratio R calculation are suppressed simultaneously. Optical influences, which occur during engine measurements, are neglected. Figure 2-20 displays three calculated fitting functions as a function of the inverted temperature. Using Planck's law with Wien's formula, the theoretical function for the signal ratio considering constant filter and IRO specifications is fitted to the measured calibration curve (Planck fit). In comparison to that, a logarithmic fit (ln fit) is displayed calculated from a three-point-calibration. Taken the calibration measurements into account, the temperature range with nearly constant gradient of signal ratio is chosen of all three measurements. Hence, at these three points, a logarithmic function is fitted and extrapolated. The third fit is a linear fit of the three calibration points. Calculating the logarithm of all signal ratios gives information about the degree of fitting. In the medium temperature range, all three fittings show quite good agreement to the calibration measurements like expected. Towards lower temperatures, the Planck fit diverges more and more to higher ratios from the measurement ratios. The linear fit shows the highest correlation. Towards higher temperature, the Planck fit crosses the measurement ratios, while the linear fit adapts quite well the measurement ratio curve. The ln fit shows a better correlation than the Planck fit, but does not adapt the measurement ratio curve as well as the linear fit.

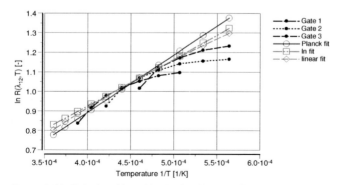

Figure 2-20: Calculated logarithmic signal ratio ln R as a function of inverted temperature for three different IRO gates and three fitting functions. Records of tungsten ribbon lamp, intensity frame averaged.

Anyway, calculating the linear regressions and their determination coefficients, all fits result in good determination coefficients of $R^2 > 0.985$ like displayed in Figure 2-21.

Furthermore, the influence of the fitting on temperature calculation can be observed. Using Planck or the ln fit results in very similar regression lines and therewith calibration constants. The use of the linear fit with its quite good adaptation to the calibration measurements yields to a slighty reduced gradient of the regression line.

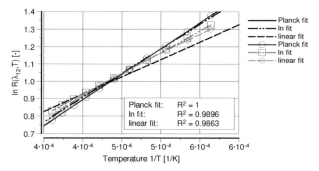

Figure 2-21: Linear regression lines for the three signal ratio fittings and their determination coefficient.

For evaluating the fittings influence on temperature calculation, Figure 2-22 displays a temperature calculation based on on-engine measurements of combustion during stratified engine operation. Based on the soot luminosity detected, its temperature is calculated using the three discussed calibration functions. Due to the quite similar regression line, no noticeable differences between the Planck fit and the ln fit can be observed. But even when using the linear fit with a slightly decreased gradient of the regression line, the temperature distribution does not differ considerably.

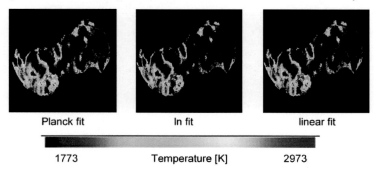

Figure 2-22: Calculated non-premixed flame temperature distribution using the three signal ratio fittings. n_{mot} = 1500 min^{-1}, imep = 4 bar, Ref-O_2-03, λ = 2.8, stratified injection, φ = 2 °CA b. TDC

Hence, the Planck fit is used for further evaluations due to its agreement with the theoretical basics of Planck's law and Wien's formula. Consequently, the calibration values A_{12} and $\ln(C_m c_{12})$ are determined using records of the tungsten ribbon lamp and fitted to the optical properties of the experimental setup.

Due to the transition of an off-engine calibration under laboratorical conditions to on-engine measurements, several effects are neglected. Like discussed detailly by di Stasio et al. [168], the influences of in-cylinder gas conditions on light signal interferences due to changing imaging wavelength are neglected as well as detailed optical properties of the glewing solid bodies, i.e. soot particles. Using the ratio of two wavelengths close to each other, these tampering effects can be suppressed to a certain degree. Finally, a three dimensional flame is imaged on a two dimensional focus plane leading to a virtual temperature distribution. Hence, information about the absolute temperature is diminished. Nevertheless, the influence of combustion modulation on the non-premixed flame temperature can be displayed and discussed. Because of the used narrow infra red filters, non-premixed combustion can be separated and qualitative impressions of soot fraction in the image area can be given.

The endoscopic access

For application of imaging measurement tools an optical access to the combustion chamber is necessary. Several designs are well established for fundamental research investigations [169]. Using transparent pistons, combustion chambers on entire or single cylinder engines allow optimum optical access and the application of complex measurement techniques. Especially quantitative measurement techniques deliver quite good results on such designs. Regarding the transferability of results, these designs lead to some strong restrictions. Changes of wall heat transfer, realistic gas exchange behaviour and thermal conditions limit the significance for lifelike engine operation. Endoscopic accesses are a compromise between optical information and real engine conditions. For imaging combustion luminosity and determining flame temperature, an optical access with high transmission and minimum changes of combustion chamber geometry is advantageous. Thus, an endoscopic access is applied to the test engine.

To gain a maximum angular field, a plane/concave lense is mounted flushly to the chamber roof using a three sleeve design with pressure air cooling (details can be found in the appendix). Figure 2-23 shows the CAD model of cylinder head segment with the endoscopical access on cylinder one. The endoscope is aligned towards the injector axis. For maximum angular field it is tilted horizontally. Due to this spark plug and injector nozzle pike can be located by the camera view at its top (compare Figure 2-24). The camera system is focused on the lens. To record the complete image information provided by the lens, distance ringes are mounted between the optical objective and the aperture of the double image optic. By this, a narrow optical beam is created for transporting the image information. The camera system is disconnected mechanically from the engine. During recording, no influences of engine shaking on the images could be noticed.

Figure 2-23: CAD model of cylinder head segment with endoscopical access and camera view into the combustion chamber

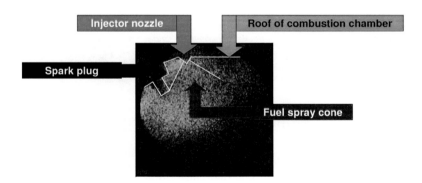

Figure 2-24: Endoscopic view into combustion chamber with visible edges.

The influence of cyclic variations on optical combustion analysis

The spark ignited gasoline engine is characterized by a certain amount of cyclic variations. These are caused by the explicit dependence of flame propagation on the induced turbulence in the combustion chamber. Depending on stochastic variations of in cylinder flow and the induced turbulent variation intensity, ignition of the air/fuel mixture and flame propagation through it varies. When stratifying the mixture, these stochastic variations even vary the stratification and therewith flame propagation and heat release. During inflammation, most influencing thermodynamic parameters are pressure and temperature as well as the residual gas mass fraction. An increasing cylinder pressure and temperature stabilizes the ignition and combustion process because the probability of creating a flame kernel is increased. On the other hand, an increasing residual gas mass fraction increases the danger of flame extinction due to the increased probability of flame quenching.

Due to their direct link to combustion and flame propagation, cyclic variations greatly influence the optically determined flame temperature. Figure 2-25 shows the temperature calculation based on a record sequence with six images taken per crank angle degree. The consecutively recorded combustion luminosity shows variations from image to image, ergo cycle to cycle. Although a tendency with higher luminosity concentration towards the intake side can be seen, still the concentration and with it the determined temperature distribution varies.

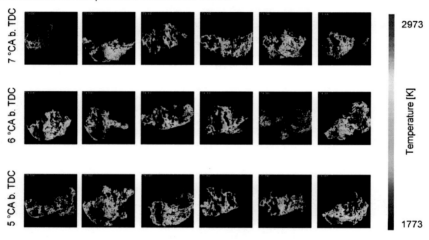

Figure 2-25: Two-dimensional T_{npf}-distribution calculated for every image of a record sequence with six images per crank angle degree. $n_{mot} = 1500\ min^{-1}$, imep = 4 bar, λ = 2.8, EGR = 0%, stratified injection

Calculating the single frame averaged temperature for every taken frame, the standard deviation of the flame temperature is calculated and its accuracy can be estimated. Figure 2-26 shows the calculated single frame averaged temperature as a function of crank angle degree (symbols) as well as the averaged temperature T_{npf}. In comparison, the zero-dimensional calculated temperature of the burned gas $T_{burned\ gas}$, based on the recorded pressure traces, is displayed.

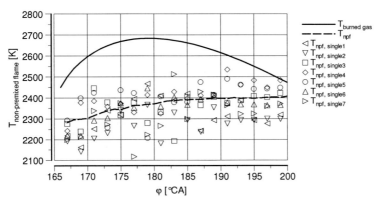

Figure 2-26: Image averaged temperatur T_{npf} and zero-dimensional temperature of the burned zone $T_{burned\ gas}$ as a function of crank angle degree; Symbols: Single image temperature, Dotted line: Frame averaged temperature. n_{mot} = 1500 min^{-1}, imep = 4 bar, λ = 2.8, EGR = 0%, stratified injection

At this operation point, the calculated single frame averaged temperature varies at maximum about 200 K similar to measurements in [160]. Considering its convergence towards the frame number taken per crank angle degree (comp. Figure 2-27, left), it can be observed that seven frames per time unit result in a trustworthable averaged temperature determination.

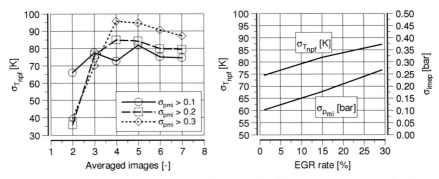

Figure 2-27: Standard deviation of calculated non-premixed flame temperature as a function of the averaged images per crank angle degree (left) and as a function of the EGR rate as well as the standard deviation of the indicated mean pressure as a function of the EGR rate.

At standard deviations of σ_{imep} > 0.3 bar, convergence is still trustworthy. The standard deviation of non-premixed flame temperature correlates quite well with the standard deviation of the cylinder pressure (comp. Figure 2-27, right). Referring to the zero-dimensional calculated temperature of the burned gas, a steady offset of 200-300 K can be observed in comparison to the optically determined flame temperature. The temperature of the burned zone can be observed as maximum gas temperature for the existing thermodynamic condition (comp. chapter 2.1.3). Due to the assumption of a stoichiometrical burned gas and the neglection of heat exchange between the cylinder zones, the temperature of the burned zone is quite similar to the adiabatic flame temperature.

2.6 Optical combustion characteristics of spray-guided stratified mixture formation

The combustion of stratified mixture is characterized by its mixture formation and the subsequent flame propagation. An optical characterization of the combustion is essential for discussion of engineering challenges. To describe the flame kernel formation and the subsequent primary flame propagation, an optical fibre spark plug sensor is used. Up to eighty optical fibres are mounted into a spark plug clockwise around the electrodes. Using saphir windows, the view of these optical fibres is aligned radially as well as vertically on four levels (comp. appendix I). Using a threshold function, the time gap between ignition and first light passage on each optical fibre is determined. The primary flame propagation can be expressed using the flame propagation speed $s_{f, prop}$

$$s_{f, prop} = s_{f, prop} \left[\frac{m}{s} \right] \cdot \left(\frac{n_{mot}}{60} \cdot 360 \right)^{-1} \cdot 10^3 \quad [mm/^\circ CA] \qquad [2\text{-}43]$$

when implying the geometrical information of the spark plug sensor and the actual engine speed and plotted as polar diagram with $s_{f, prop}$ as radius and the radial sensor view angle as polar angle [170]. Figure 2-28 shows the polar diagram of the flame propagation speed for all four view levels for the reference standard operation point. For easier recognition, the inlet side is always on six o' clock.

Reminding the slightly excentric position of the spark plug towards exhaust, a distinctive drift of the flame kernel towards the exhaust side can be observed. Flame propagation is mainly controlled by the mixture structure.

Figure 2-29 displays the primary flame propagation during flame kernel formation as polar diagram for two exhaust gas air-/fuel ratios. Like observed before, the flame kernel drifts towards the exhaust. With decreasing exhaust gas air-/fuel ratio, flame propagation speed $s_{f, prop}$ decreases due to increased residual gas fraction and becomes more even distribution in the radial direction.

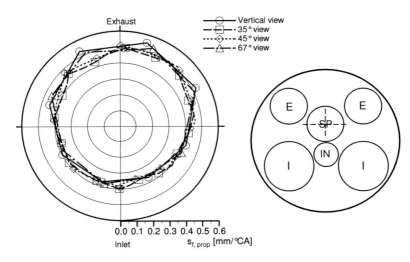

Figure 2-28: Primary flame propagation during inflammation determined by optical fibre spark plug sensor (all views). n_{mot} = 1500 min^{-1}, imep = 4 bar, Ref-O$_2$-03, λ = 2.8, stratified injection

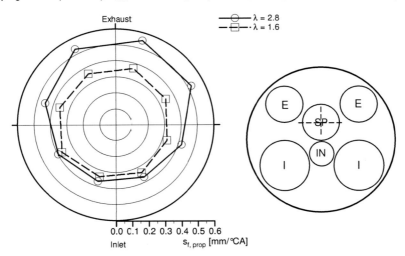

Figure 2-29: The influence of exhaust gas air-/fuel ratio on primary flame propagation determined using optical fibre spark plug sensor (vertical view). n_{mot} = 1500 min^{-1}, imep = 4 bar, Ref-O$_2$-03, stratified injection, Cyl. 1

Using an intensity pattern, measured light intensities are averaged in terms of view levels and cylinder sections. By this, qualitative information about the spatial occurrence of combustion phenomena can be gained. Figure 2-30 shows an intensity pattern for two exhaust gas air-/fuel ratios. Intensities are averaged for four cylinder sections like displayed on the right in Figure 2-30. All intensities develop with an increase at 15 °CA b. TDC, reach an early maximum and decrease continuously similar to a rate of heat release.

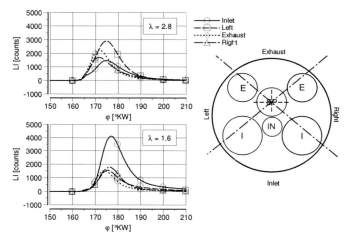

Figure 2-30: Intensity pattern for two exhaust gas air-/fuel ratios by optical fibre spark plug sensor. n_{mot} = 1500 min^{-1}, imep = 4 bar, Ref-O_2-03, stratified injection, Cyl. 1

At λ = 2.8, intensity develops quite even radially. Towards the left side, a slightly increased intensity maximum can be observed. With decreased air excess, a distinctive increased intensity maximum on the inlet side can be found. Intensity is more than twice as high as in the other cylinder sections. Still all curves show a continous developing over time. Thus, threshold analysis shows a drift of the flame kernel towards the exhaust, while increased light intensity can be found towards the inlet with decreasing exhaust gas air-/fuel ratio.

To gain spatial information about mixture formation and subsequent combustion, CFD analysis are performed. Analogue to Figure 2-8, Figure 2-31 displays the CFD calculated fuel mass fraction as an intersection during a triple injection under strong enriched mixture conditions. This global understoichiometrical mixture condition is chosen to generate a maximum amount of soot luminosity. Due to the spray characteristic, enriched fuel vapour rings are composed located below the spark plug and stabilized by the slight piston bowl. To stabilize ignition, the third injection enriches the mixture at the spark plug and the stratified mixture is ignited at the exhaust side of the mixture rings.

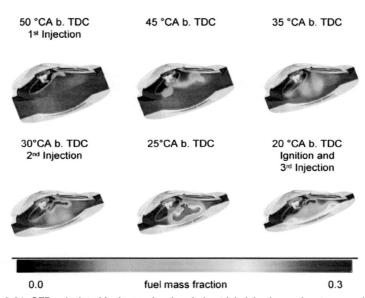

50 °CA b. TDC   45 °CA b. TDC   35 °CA b. TDC
1st Injection

30°CA b. TDC   25°CA b. TDC   20 °CA b. TDC
2nd Injection   Ignition and
                               3rd Injection

0.0   fuel mass fraction   0.3

Figure 2-31: CFD calculated fuel mass fraction during triple injection under strong enriched engine mixture condition. *Intersection at spark plug / injector; n_{mot} = 1500 min^{-1}, imep = 4 bar, λ < 1, stratified injection*

This carburetion characteristic results in a certain distribution of soot fraction and with it in a characteristical distribution of non-premixed flame temperature. This is qualitatively confirmed by imaging records of stratified combustion and corresponding CFD calculations. Figure 2-32 shows the temperature distribution calculated by CFD and 2-CRP during stratified combustion 20 °CA a. TDC.

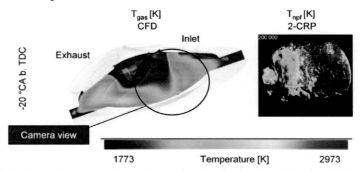

T_{gas} [K] T_{npf} [K]
CFD   2-CRP

Inlet

Exhaust

-20 °CA b. TDC

Camera view

1773   Temperature [K]   2973

Figure 2-32: Comparison of calculated temperature distribution via CFD and 2-CRP. *n_{mot} = 1500 min^{-1}, imep = 4 bar, λ < 1, 20 °CA a. TDC, i-C8, stratified injection*

Both images indicate a hot zone left of the spark plug, i.e. towards the exhaust. In the middle below the injector, quite low temperatures can be observed because of a fuel deficit due to the spray characteristic. Below the inlet valves again, higher temperatures can be found originated from conversion of conditioned air-/fuel mixture, the inlet side of the composed spray ring. Here, temperatures up to 2700 °C can be found. Figure 2-34 displays the CFD calculated burned gas temperature as well as the calculated non-premixed flame temperature distributions using averaged images for the heat release period. From the first image on, a zone with increased temperature can be observed towards the inlet. This zone remains stable for the whole image sequence in the images as well as when looking at the CFD calculated distribution. Here, an enriched zone is composed during carburetion, while a leaner zone with low flame temperature below the injector can be found, which is characteristic for this combustion process and can be observed throughout all image records. At the exhaust side of the CFD calculations, a second hot zone can be observed originated from the ignited mixture ring at the spark plug which is decentralized towards the exhaust (compare cylinder scetch in Figure 2-30). From there on, the hot zone moves towards the exhaust side during the piston descends expressing the main flame propagation. According to the triple flame theory, after this primary flame propagation non-premixed flame occur generating soot particles, which can be detected using 2-CRP. This corresponds with the found intensity pattern results in Figure 2-30 as well as in Figure 2-33. Figure 2-33 shows the intensity curves averaged for four prior defined cylinder sections at low exhaust gas air-/fuel ratio using iso-octane. Like observed in Figure 2-30 with reference fuel, a twice as high increased light intensity towards the inlet is found. The distinctively increased intensity maximum towards the inlet is directly related to the high soot luminosity with high non-premixed flame temperature towards the inlet.

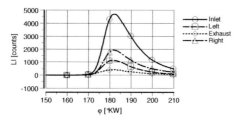

Figure 2-33: Intensity pattern by optical fibre spark plug sensor. n_{mot} = 1500 min^{-1}, imep = 4 bar, i-C8, λ = 1.4, stratified injection, Cyl.1

Hence, primary flame propagation is mainly exhaust oriented due to mixture ring location around the spark plug. The inlet oriented mixture burns under formation of soot and non-premixed flames.

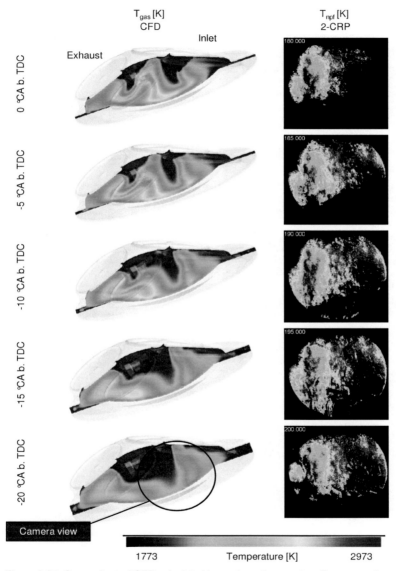

Figure 2-34: Comparison of CFD calculated burned gas temperature T_{gas} versus the calculated non-premixed flame temperature T_{npf}.
$n_{mot} = 1500\ min^{-1}$, imep = 4 bar, $\lambda < 1$, i-C8, stratified injection

This is verified by its local stability in the inlet hot zone and therewith approves the diffusion controlled combustion process in comparison to the turbulence influenced combustion with more premixed conditions at the exhaust side. Evenmore, optical analysis of the combustion of stratified mixture verifies the phenomena of partially-premixed combustion, which is controlled by the carburetion.

Quite remarkable, the temperature distribution increases to the center of the hot zone, where no light emission signal of solid bodies, i.e. soot particles, can be observed. This correlates qualitatively with measurements presented by Stojkovic et al. [171]. Flame temperature increases above the temperature limit for soot particle formation (compare Figure 2-13). Formed particles are too small to emit recordable light or are pyrolized. Like mentioned before, the displayed temperature is a calculated surface temperature of formatted soot particles in the non-premixed flame. Hence, a statistical correlation between detected light and theorectical soot distribution can be assumed.

Figure 2-35 shows a statistical evaluation of the temperature distribution ρ_{Tnpf}. The temperature distribution of every image, i.e. crank angle position, with valid temperature information is evaluated. For depiction, a crank angle averaged temperature distribution is calculated (ρ_{Tnpf} CA-avg.) as well as the hull for all evaluated temperature distributions (ρ_{Tnpf} Max/Min). Non-premixed flame Temperature distribution can be described as Gauss' normal distribution with its maximum at around 2400 K. Towards higher temperature, the evaluation is limited due to the measurement evaluation technology. Pyrolysis of the glewing soot particles is expected giving no evaluatable light signal ratios.

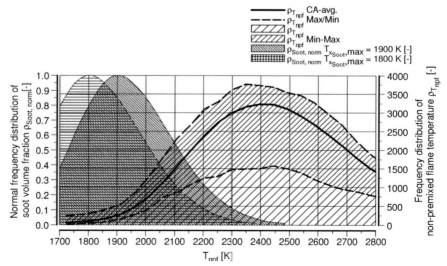

Figure 2-35: Crank angle averaged ($\varphi = 165 \div 200$ °CA) frequency distribution of non-premixed flame temperature ρ_{Tnpf} and assumed as well as normed normal frequency distribution of soot volume fraction $\rho_{Soot, norm}$ as a function of non-premixed flame temperature. $n_{mot} = 1500$ min^{-1}, imep = 4 bar, i-C8, stratified injection, Cyl.1

Additonally in Figure 2-35, an assumed normed frequency distribution of soot volume fraction $\rho_{Soot,norm}$ describing a Gauss' normal distribution function [172] is displayed for two different soot volume fraction maximum (1800 K and 1900 K). It can be observed, that the temperature distribution, describing the number of pixels (i.e. glewing particles in the flame) with a concurrent temperature, is shifted to higher temperatures in comparison to the assumed soot volume fraction distribution. Hence, two intersections of the evaluated temperature distribution and the assumed two soot volume fraction distributions can be recognized. Theoretically, if every pixel with its concurrent temperature represents the corresponding soot volume fraction, it could be concluded, that:

- in areas with a non-premixed flame temperature of around 2100 K the highest soot concentration can be found,

- a shift of the soot distribution to higher temperatures results in increasing soot concentration and

- areas with non-premixed flame temperatures higher than 2400 ÷ 2500 K cannot be linked to a soot volume fraction.

Thus, particle surface temperature, which is calculated based on the measured light intensity, is further increased in comparison to the theoretically gas temperature for soot formation. Taking the contemporaneous soot oxidation into account (comp. Chapter 2.4.1), the higher surface temperature can be reasoned by the exothermic character of soot oxidation.

If the soot volume fraction maximum is shifted to higher temperatures, due to decreasing pressure for example [172], a distinctive increase of the intersection between soot volume fraction and non-premixed flame temperature during combustion can be assumed according to the growing intersection shown in Figure 2-35 for two different soot fraction maximum temperatures.

Summary

Optical analysis of combustion phenomena of a combustion process with mixture stratification using optical fibre spark plug as well as the introduced 2-CRP verifies the model approach of partial-premixed combustion for describing combustion. It should be pointed out that

- zones of non-premixed flame occurrence and soot formation are strongly depending on mixture formation and are local stable and

- combustion in stratified DISI engines is "carburetion controlled".

Regarding the determined non-premixed flame temperature, it has to be remarked, that, due to the physical approach of the measurement system, the surface temperature of the glewing particles is determined. Under theoretical considerations of temperature ranges of soot formation and oxidation, it can be assumed that this surface temperature is influenced by exothermic reactions during particle oxidation.

3 Study on combustion characteristics of turbo charged direct injection SI-engines

Spray-guided DISI engine combustion characteristics are dominated by partial-premixed combustion phenomena at part-load stratified mixture and knock phenomena at full-load conditions. Due to a partial-premixed combustion, the occurrence of non-premixed combustion phases is unevitable and needs to be understood because of the danger of increased soot emission. While knock phenomena do not differ from already described knock mechanisms [31], the influence of unburned or partially burned exhaust gas components on autoignition is investigated.

3.1 Combustion characteristics of spray-guided stratified mixture formation

Caused by the physical and chemical processes of a partial-premixed combustion when stratifying the air/fuel mixture, the combustion is sensibly influenced by engine parameters. Using turbo charging as well as external cooled EGR and varying engine parameters like charge motion flaps lead to interacting changes of the engine state. Even more influence on the combustion process during stratification can be seen when using complex multiple injection strategies for creating optimal mixture stratification. The interaction of the injected spray with the in-cylinder gas and the resulting combustion mixture can be varied in regards to air-/fuel-/residual-mixture and its distribution as well as its degree of carburetion. The caloric properties of the burning mixture influence the temperature and rate of heat release, which are directly influencing the production and reduction of unwanted pollutant emissions during combustion. Therefore, next to the variation of carburetion via injection strategies and rail pressure, the influence of the caloric properties of the combustion gas on flame temperature and the occurrence of non-premixed combustion is targeted. Varying the amount of external recirculated exhaust gas and/or fresh air, caloric properties of the intake gas are modified. Via a fuel study, the influence of ethanol on non-premixed combustion and its temperature is described. Due to its oxidation enhancing alcohol group, the formation and oxidation of soot during combustion will be influenced. Furthermore, the modified caloric properties of the resulting burned gas will be considered regarding their influence of pollutant emissions. Next to primary reference fuel modifications, commercial multi-component fuels with similar caloric properties are compared. Test bench measurements take energy consumption and exhaust gas emissions into account.

As standard operation point for stratified combustion during engine run, an engine speed of n_{mot} = 1500 min^{-1} and an engine load of imep = 4 bar are defined (more details in appendix b). Due to ignition stability, ignition angle and injection timing need to be constant. A change of the gap between ignition angle and injection timing results in misfire occurrence. Using the two-colour ratio pyrometry, non-premixed combustion luminosity is visualized and the flame temperatue T_{npf} is determined. Detailed thermodynamic analysis (0-D DTDA) based on cylinder pressure measurements describes heat release, caloric properties, the resulting zero-dimensional gas temperatues and the occurring losses. An optical fibre spark plug sensor gives informations about primary flame propagation during inflammation and the qualitative occurrence of combustion luminosity. Comparing CFD calculations give detailed information about the carburetion and the resulting mixture distribution during the high-pressure cycle. Like implicated before, combustion process is influenced by several engine parameters. Thus, the general influence of engine and fuel parameters on heat release and their implications on non-premixed combustion and non-premixed flame temperature should be discussed.

3.1.1 Effect of tumble flap on non-premixed flame temperature

Closing tumble flap results in modification of the in-cylinder flow. With regard to engine run, closing tumble flap to more than fifty percent leads to decreased engine efficiency like shown in Figure 3-1.

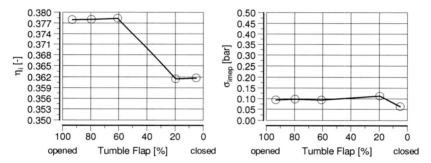

Figure 3-1: Indicated engine efficiency and standard deviation of indicated mean pressure as a function of tumble flap position. n_{mot} = 1500 min^{-1}, imep = 4 bar, Ref-O$_2$-03, IGA = 26.25 °CA b. TDC, stratified injection

This involves a slight increase of cyclic variations up to twenty percent closed tumble flap. At complete closed tumble flap, cyclic variations are lower (Figure 3-1, right diagram). Despite an improvement of combustion stability, engine efficiency decreases with closed tumble flap because of increased pumping losses which can be observed in Figure 3-2, right diagram. Additionally, the increased flow losses in the intake reduce the achievable air-/fuel ratio reducing the caloric efficiency. The tumble flap blocks the lower inlet port section completely. The in-streaming air streams over the flap through a reduced cross section resulting in a more aligned cylinder inflow with increased stream velocity and turbulence. Because of this, the cylinder charge is more ignitable resulting in a more stable and faster inflammation. The difference between a complete closed and a not complete closed tumble flap gets more evident looking at the calculated time of fifty percent mass fuel burned for different tumble flap positions in Figure 3-2.

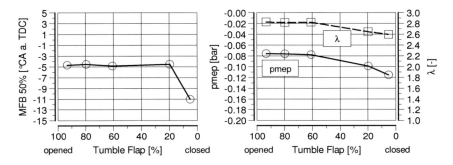

Figure 3-2: Time of fifty percent mass fuel burned, indicated mean pressure during gas exchange and air-/fuel ratio of exhaust gas as a function of tumble flap position.
n_{mot} = 1500 min^{-1}, imep = 4 bar, Ref-O_2-03, IGA = 26.25 °CA b. TDC, stratified injection

Up to still twenty percent tumble flap opened, MFB 50% remains constant. With complete closed tumble flap, MFB 50% is considerably shifted to earlier time as an indication for much faster heat release reasoned by the increased turbulent kinetic energy induced by the complete closed tumble flap. Explaining it the other way around, a not complete closed tumble flap results in changes of the in cylinder flow. But due to a still present air stream in the lower intake section, there is no noticeable increase of turbulence.

Figure 3-3 shows the cylinder pressure trace and mass averaged gas temperature versus crank angle degree during heat release for three tumble flap positions at standard operation point. Figure 3-4 shows the rate of heat release and burn function for three tumble flap positions at standard operation point.

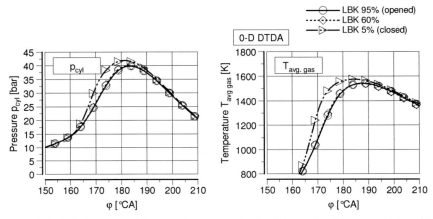

Figure 3-3: Cylinder pressure trace and mass averaged gas temperature versus crank angle degree during heat release for three tumble flap positions. n_{mot} = 1500 min^{-1}, imep = 4 bar, Ref-O_2-03, IGA = 26.25 °CA b. TDC, stratified injection, Cyl.1

Like observed in Figure 3-2, only at closed position the rate of heat release reacts. The gradient is stronger, the maximum higher and, consequently, heat release takes place earlier compared to the open and sixty percent closed tumble flap position. Hence, maximum pressure and mass averaged gas temperature are higher. This indicates that only at closed position a certain rise of turbulent kinetic energy is introduced into the combustion chamber accelerating heat release. Towards the end of heat release, heat release remains higher with closed tumble flap.

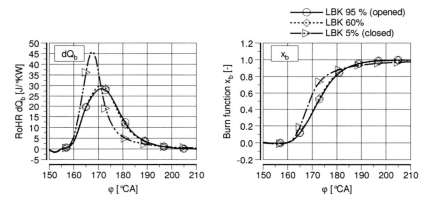

Figure 3-4: Rate of heat release and burn function versus crank angle degree for three tumble flap positions. n_{mot} = 1500 min^{-1}, imep = 4 bar, Ref-O$_2$-03, IGA = 26.25 °CA b. TDC, stratified injection, Cyl.1

Referring to the flame kernel formation and the subsequent primary flame propagation, closing the tumble flap leads to a slight deformation of the known flame propagation shape in the polar diagram in Figure 3-5. With closed tumble flap (LBK = 5 %), the primary flame propagation moves towards the right. Next to the main drift towards the exhaust, flame propagation towards the inlet is accelerated corresponding with the decreased ignition delay and faster heat release.

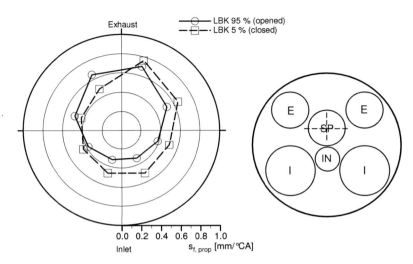

Figure 3-5: The influence of tumble flap position on primary flame propagation determined by optical fibre spark plug sensor (vertical view). n_{mot} = 1500 min^{-1}, imep = 4 bar, Ref-O_2-03, IGA = 26.25 °CA b. TDC, stratified injection, Cyl.1

Figure 3-6 displays the temperature of non-premixed flame (left) and the temperature of burned gas for the minimum and maximum tumble flap positions. The diagrams are limited to the crank angle range of trustworthy temperature calculation of the 2-CRP. The optical calculated flame temperature T_{npf} shows a quite good correlation to the zero-dimensional gas temperature. At open tumple flap, the temperature curves are similar with a maximum around top dead center (TDC = 180 °CA). With complete closed tumble flap, the temperature maximum is shifted to earlier position correlating to the faster heat release (compare Figure 3-4). The maximum of optical calculated temperature is quite equal to the opened tumble flap positions. Looking at the zero-dimensional temperature, the closed tumble flap position results in a decreased temperature maximum compared to the opened position. After the temperature maximum is reached, zero-dimensional gas temperature decreases distinctively while the non-premixed flame temperature remains on a similar level caused by the thermal inertia of the light emitting soot particles.

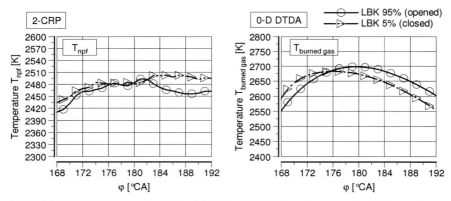

Figure 3-6: Temperature of non-premixed flame and burned gas temperature versus crank angle degree for two tumble flap positions. n_{mot} = 1500 min^{-1}, imep = 4 bar, Ref-O_2-03, IGA = 26.25 °CA b. TDC, stratified injection, Cyl.1

Due to an increased pressure difference between intake and exhaust at closed tumble flap, an increased residual gas fraction and reduced air-/fuel ratio leads to lowered burned gas temperature in opposition to the mass averaged gas temperature in Figure 3-3.

The measured exhaust gas emissions in Figure 3-7 and Figure 3-8 express the influence of the tumble flap on heat release and combustion temperature. HC emissions correlate with the measured cyclic variations in Figure 3-1. At closed tumble flap position, less flame quenching can be expected due to a short ignition delay and a fast heat release. Although no increase of maximum flame temperature at closed tumble flap can be found, the earlier rise leads to an improved oxidation of composed CO during combustion due to increased available oxidation energy and therewith decreased CO emission.

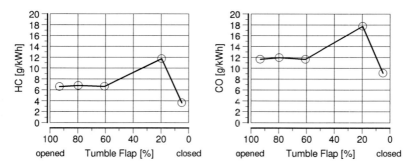

Figure 3-7: Specific raw emissions of unburned hydrocarbons HC and carbon monoxide CO as a function of tumble flap position. n_{mot} = 1500 min^{-1}, imep = 4 bar, Ref-O_2-03, IGA = 26.25 °CA b. TDC, stratified injection

The detected NO_x emissions in Figure 3-8 show contrair behaviour with a strong increase at closed tumble flap position corresponding to the development of MFB 50%. Due to earlier combustion and the resulting earlier rise of combustion temperature, considerable more NO_x is composed. Looking at the detected smoke emissions, a marginal maximum with slightly closed tumble flap can be observed correlating with slightly increased HC emissions. This indicates a more inhomogenous carburetion. With further closing tumble flap, smoke emissions close to the detection limit are found, Figure 3-8 right diagram.

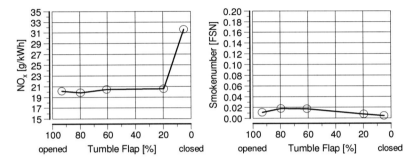

Figure 3-8: Specific raw emissions of nitric oxide and smoke emission as a function of tumble flap position. $n_{mot} = 1500\ min^{-1}$, imep = 4 bar, Ref-O_2-03, IGA = 26.25 °CA b. TDC, stratified injection

3.1.2 Effect of fuel rail pressure on non-premixed flame temperature

Carburetion and the resulting air-/fuel mixture can be influenced by modification of droplet size and the consequential evaporation process as well as by the timing and mass weighting of multiple injection pulses. While injection timing and mass weighting is directly controlled, droplet size modifications are introduced by modifications of the fuel stream when injected into the combustion chamber. Primarily, this is done by the choice of the injector type and its geometry like hole shape and diameter. During engine operation, the droplet size can be modified by varying the fuel pressure. Under assumption of a constant geometry and back pressure, varying the fuel pressure yields to changed droplet sizes [31] due to a changed spray impulse. Hence, the evaporation process as well as the spray penetration is modified and with it the mixture condition at ignition time.

Figure 3-9 shows the Cylinder pressure trace and mass averaged gas temperature versus crank angle degree during heat release for two fuel rail pressures. At higher fuel rail pressure an earlier pressure rise indicating a shorter ignition delay can be found. This results in a higher mass averaged gas temperature.

64

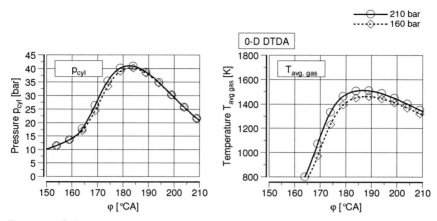

Figure 3-9: Cylinder pressure trace and mass averaged gas temperature versus crank angle degree during heat release for two fuel rail pressures. n_{mot} = 1500 min^{-1}, imep = 4 bar, Ref-O_2-03, λ = 2.8, IGA = 26.25 °CA b. TDC, stratified injection, Cyl.1

Looking at the calculated rate of heat release and the corresponding burn function in Figure 3-10, a corresponding earlier heat release rate with increasing rail pressure can be observed.

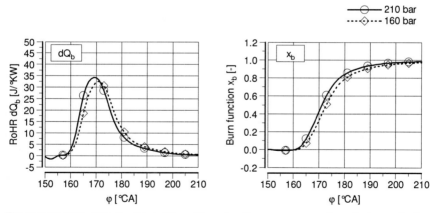

Figure 3-10: Rate of heat release dQ_b and burn function x_b versus crank angle degree for two fuel rail pressures. n_{mot} = 1500 min^{-1}, imep = 4 bar, Ref-O_2-03, λ = 2.8, stratified injection, Cyl.1

This is veryfied by optical analysis of primary flame propagation using the optical fibre spark plug sensor in Figure 3-11. During flame kernel formation, primary flame front velocity $s_{f,\ prop}$ increases with increasing rail pressure. Compared with the discussed burn function in Figure 3-10, the increase of primary flame front velocity is higher.

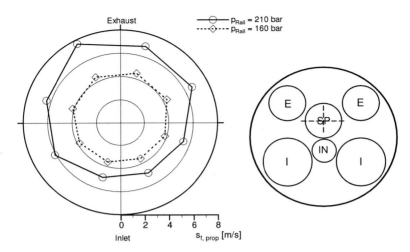

Figure 3-11: The influence of rail pressure on primary flame propagation determined using optical fibre spark plug sensor (vertical view). n_{mot} = 1500 min^{-1}, imep = 4 bar, Ref-O_2-03, λ = 2.8, stratified injection, Cyl.1

Thus, a shortened ignition delay can be assumed resulting in an increased primary flame propagation velocity and an earlier heat release. The flame front propagation shows an increasing drift component towards the exhaust.

The earlier heat release results in earlier rises of combustion temperature as can be found in Figure 3-12 corresponding to the earlier times of maximum heat release and the accelerated primary flame propagation.

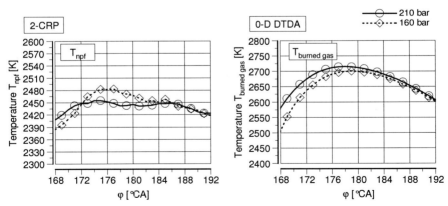

Figure 3-12: Temperature of non-premixed flame and burned gas temperature versus crank angle degree for four fuel rail pressures. n_{mot} = 1500 min^{-1}, imep = 4 bar, Ref-O_2-03, λ = 2.8, stratified injection, Cyl.1

Comparing the calculated non-premixed flame temperature in the left diagram, no clear difference in the maximum temperature can be found. Independent of the rail pressure a flame temperature of roundabout 2450 K is observed, while the zero-dimensional temperature of burned gas shows an increase in burned gas temperature in agreement with mass averaged gas temperature shown in Figure 3-9.

Referring to the influence of rail pressure on droplet size, spray penetration and carburetion, Figure 3-13 shows a continous decrease of HC emissions with increasing rail pressure. With increasing rail pressure the droplet size decreases. Hence, less time is needed for breaking up of the droplet and the subsequent evaporation [31],[173]. The propabiblity of not yet complete evaporated fuel decreases as well as the probability of fuel droplets penetrating too far into the combustion chamber. They can be reached early enough by the flame front before flame quenching occurs [57]. As a further indicator for improved carburetion with increasing rail pressure, CO emissions decrease similary with increasing rail pressure. Due to more enleaned zones in the combustion chamber, less CO is composed locally. The accelerated heat release results in higher combustion temperature, which leads to higher CO oxidation rate.

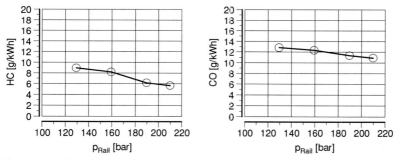

Figure 3-13: Specific raw emissions of unburned hydrocarbons HC and carbon monoxide CO as a function of fuel rail pressure. $n_{mot} = 1500\ min^{-1}$, imep = 4 bar, Ref-O_2-03, $\lambda = 2.8$, stratified injection

With increasing rail pressure and the resulting continous acceleration of heat release, NO_x emissions increase quite linearly (Figure 3-14, left diagrams). Due to the improved evaporation process, local air-/fuel ratio is expected to be shifted to more lean conditions enhancing the formation of NO_x. Still the formation of CO and NO_x is mainly influenced by the earlier combustion phasing (Figure 3-14, lower diagram) and the resulting increased combustion temperature (compare Figure 3-9).

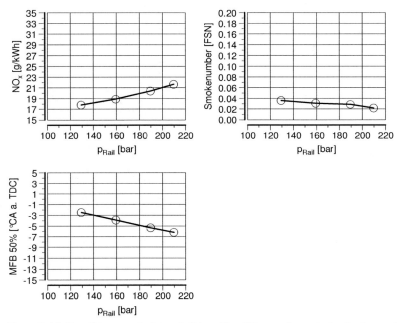

Figure 3-14: Specific raw emissions of nitric oxides NOx, smoke emission (entire engine) and time of fifty percent mass fuel burned as a function of fuel rail pressure. n_{mot} = 1500 min^{-1}, imep = 4 bar, Ref-O_2-03, λ = 2.8, stratified injection

Corresponding to that, with increasing rail pressure slightly decreased smoke emissions can be found in Figure 3-14, right diagram. The decreased combustion luminosity when using 2-CRP in Figure 3-15 verifies the soot reducing effect of increased rail pressure. At crank angles position before top dead center, the distribution of camera pixels detecting soot luminosity and giving temperature information is quite independent of rail pressure. After top dead center corresponding to later than eighty percent mass fuel burned, the image area with detected soot luminosity decreases with increasing rail pressure. Like observed in Figure 3-12, the calculated level of non-premixed temperature does not differ with increased rail pressure.

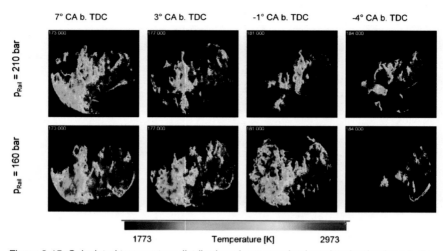

Figure 3-15: Calculated temperature distribution of non-premixed combustion for three fuel rail pressures. $n_{mot} = 1500\ min^{-1}$, imep = 4 bar, Ref-O_2-03, $\lambda = 2.8$, stratified injection, Cyl.1

Figure 3-16 shows the developing of the relative soot luminosity fraction versus crank angle degree for two fuel rail pressures. Taken from the red filtered post processed image, all pixels showing light intensity in the red shifted spectrum are counted and summed up per crank angle degree. Hence, information about area of soot lightning combustion is given.

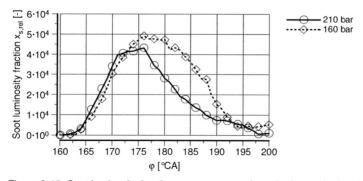

Figure 3-16: Soot luminosity fraction $x_{s,rel}$ versus crank angle degree for four fuel rail pressures. $n_{mot} = 1500\ min^{-1}$, imep = 4 bar, Ref-O_2-03, $\lambda = 2.8$, stratified injection, Cyl.1

With increasing fuel pressure the maximum of soot luminosity fraction decreases and is shifted to earlier position correlating with the maximum of heat release rate (comp. Figure 3-10). Much more noticeable, the falling edge of the soot luminosity fraction is delayed with decreasing fuel pressure. Due to decreasing carburetion quality with decreasing fuel pressure, more soot is composed. Due to the later heat release, soot oxidation is diminished [31]. Finally, the decreased fuel pressure and the linked increased droplet size give the opportunity of larger soot particles, which remain longer at higher temperature. The cooling of the composed particles is delayed and they are visible for a longer time.

3.1.3 Effect of exhaust gas air-/fuel ratio on non-premixed flame temperature

During real life engine operation, air-/fuel ratio is limited by a minimum exhaust gas temperature. Air-/fuel ratio can be controlled using the throttle in front of the intake manifold or increasing EGR rate. Reducing air-/fuel ratio at constant loads using the throttle yields to decreased engine efficiency like displayed in Figure 3-17. The standard deviation of indicated mean pressure increases degressively up to an air-/fuel ratio of 2 and then decreases.

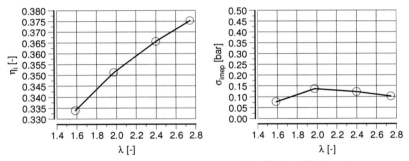

Figure 3-17: Indicated engine efficiency and standard deviation of indicated mean pressure as a function of exhaust air-/fuel ratio. n_{mot} = 1500 min^{-1}, imep = 4 bar, Ref-O_2-03, stratified injection

Major effect when throttling the engine is an increase of needed pumping work. This can be illustrated using a p-V-Diagram. Figure 3-18 shows cylinder pressure trace versus cylinder volume (p-V-Diagram) for four exhaust air-/fuel ratios. With decreasing air-/fuel ratio the gas exchange cycle gets larger. Hence, increasing pumping work is performed due to increased pumping losses.

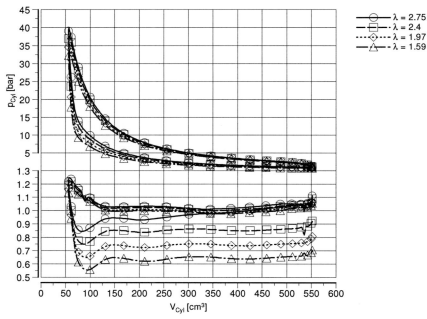

Figure 3-18: Cylinder pressure trace versus cylinder volume (p-V-Diagram) for four exhaust air-/fuel ratios. n_{mot} = *1500 min⁻¹*, imep = 4 bar, IGA = 26.25 °CA b. TDC, Ref-O_2-03, stratified injection

Consequently, an increase of high pressure cycle work is necessary resulting in increased mass averaged gas temperature due to an increase of the specific with decreasing air-/fuel ratio heat like can be observed in Figure 3-19.

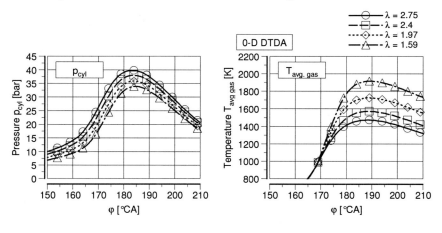

Figure 3-19: Cylinder pressure trace and mass averaged gas temperature versus crank angle degree for four exhaust air-/fuel ratios. n_{mot} = 1500 min^{-1}, imep = 4 bar, IGA = 26.25 °CA b. TDC, Ref-O_2-03, stratified injection

Evenmore, pressure and mass averaged gas temperature traces are shifted to later timing with decreasing air-/fuel ratio. This is caused by a later and lower rate of heat release like displayed in Figure 3-20. With decreasing air-/fuel ratio by reducing air mass flow via the throttle, residual gas fraction increases due to an increasing pressure difference between exhaust and inlet (compare gas exchange cycle in Figure 3-18). Hence, with decreasing air-/fuel ratio, igniton delay is getting longer and heat release is slightly shifted to later positions. This is approved by the optical analysis of the primary flame front propagation in Figure 2-29 using the optical spark plug sensor. Keeping in mind that mixture enrichment is expected to increase rate of heat release, this behavior indicates a stable stratification of the mixture.

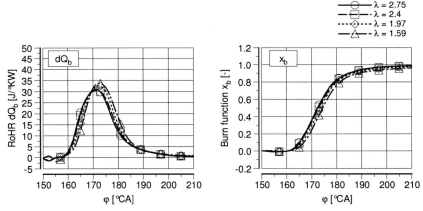

Figure 3-20: Rate of heat release dQ_b and burn function x_b versus crank angle degree for four exhaust air-/fuel ratios. n_{mot} = 1500 min^{-1}, imep = 4 bar, Ref-O_2-03, stratified injection

This indication is approved by resulting exhaust gas emissions displayed in Figure 3-21. HC emissions decrease with decreasing air-/fuel ratio. The less lean mixture decreases the probability of flame quenching. Additionally, the decreased mass flow through the engine decreases the absolute emission mass flow. CO emissions decrease with decreasing air-/fuel ratio. At high air-/fuel ratio, quite low CO concentrations are detected due to sufficient oxygen for oxidation. The decreased mass flow results in decreasing specific CO emissions. Below an air-/fuel ratio of λ = 2, CO emissions begin to increase again. The local combustion mixture is getting more and more enriched resulting in enhanced CO formation. Because of reduced air supply, CO oxidation is diminished. The effect of temperature is definitely subsidiary.

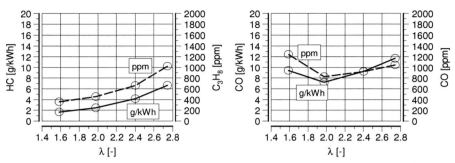

Figure 3-21: Specific raw emissions and their concentration of unburned hydrocarbons HC and carbon monoxide CO as a function of exhaust air-/fuel ratio. n_{mot} = 1500 min^{-1}, imep = 4 bar, Ref-O_2-03, stratified injection

This is supported by the developing of NOx emissions with decreasing air-/fuel ratio. According to Jakobs [56], the developing of NO_x emissions can be used to determine the local air-/fuel ratio of combustion. Looking at Figure 2-10, it can be remarked that, under homogenous mixture condition, the minimum of CO emissions as well as the maximum of NO_x emissions can be observed at an exhaust gas air-/fuel ratio of λ = 1.1. Under stratified mixture condition, these minimum and maximum can be found between an exhaust gas air-/fuel ratio around 1.6 < λ < 2.0 like in Figure 3-21 and Figure 3-22. In Figure 3-22, left diagram, NO_x emissions increase with decreasing air-/fuel ratio with a maximum between 1.6 < λ < 2.0 corresponding to the minimum of CO emissions. Hence, this indicates that strong stratification can be assumed with a burnable mixture with an air-/fuel ratio around λ = 1 and surrounding excess air.

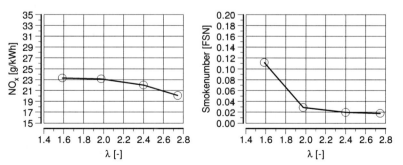

Figure 3-22: Specific raw emissions of nitric oxide and smoke emission (entire engine) as a function of exhaust air-/fuel ratio. n_{mot} = 1500 min^{-1}, imep = 4 bar, Ref-O_2-03, stratified injection,

Finally, the split of losses according to Weberbauer et al. [15] in Figure 3-23 explains the effect of decreasing air-/fuel ratio on engine efficiency.

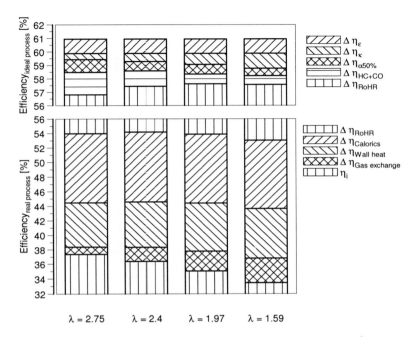

λ = 2.75   λ = 2.4   λ = 1.97   λ = 1.59

Figure 3-23: Split of losses for four exhaust air-/fuel ratios. n_{mot} = 1500 min^{-1}, imep = 4 bar, Ref-O_2-03, stratified injection

Major effect on engine efficiency decrease is the increase of pumping losses (up to sixty percent of engine efficiency decrease). Still, the increase of residual gas mass fraction and the increase of mass averaged gas temperature result in increased losses (Δ η_{RoHR} and Δ $\eta_{Wall\ heat}$) while the losses due to incomplete combustion are reduced (Δ η_{HC+CO}).

Looking at the flame temperature of non-premixed combustion in Figure 3-24, the continous retard of heat release can be retrieved in the times of temperature maxima. The temperature levels itselves do not differ with a clear trend. The calculated zero-dimensional temperature of the burned gas shows the retard of temperature maxima according to the tendential retard of heat release. A clear trend to lower combustion temperature with decreasing air-/fuel ratio can be found. This is caused by the increasing residual gas mass fraction due to throttling the engine. Since the air-/fuel ratio of the burnable mixture is assumed as stoichiometric, the decrease of the exhaust gas air-/fuel ratio results in increasing averaged gas temperature like has been observed in Figure 3-19. This assumption of stoichiometric burnable mixture with surrounding excess air is supported by the developing of CO and NO_x emissions with decreasing air-/fuel ratio in Figure 3-21 and Figure 3-22.

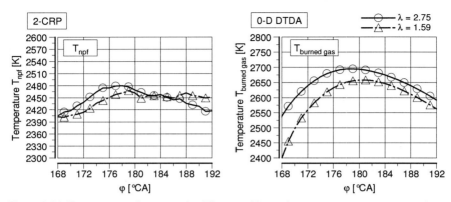

Figure 3-24: Temperature of non-premixed flame and burned gas temperature versus crank angle degree for two exhaust air-/fuel ratios. $n_{mot} = 1500\ min^{-1}$, imep = 4 bar, Ref-O_2-03, stratified injection, Cyl.1

With decreasing air-/fuel ratio, the probability of occurrence of non-premixed combustion phenomena increases resulting in increased soot formation. Figure 3-22, right diagram, shows an increase of detected smoke emissions with decreasing air supply. This is qualitatively in agreement with the soot luminosity fraction shown in Figure 3-25. Due to lens fouling, no lower exhaust gas air-/fuel ratios can be measured optically. With decreasing air-/fuel ratio the maximum of relative soot luminosity fraction is shifted to later position correlating quite well with the determined rate of heat release. A distinctive delay of the found soot luminosity fraction on the falling side of the curve for λ = 1.6 can be seen. This corresponds with the increase of smoke emissions towards lower air-/fuel ratios. Furthermore, the longer remain of glowing soot particles indicates that the decreased availability of soot oxidation is a relevant reason for increased smoke emissions.

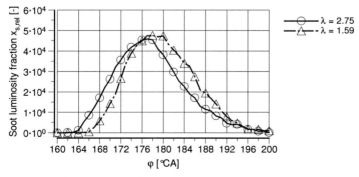

Figure 3-25: Soot luminosity fraction $x_{s,rel}$ versus crank angle degree for four exhaust gas air-/fuel ratios. $n_{mot} = 1500\ min^{-1}$, imep = 4 bar, Ref-O_2-03, stratified injection, Cyl.1

To take a global view at emissions formation, λ-T-Maps are quite significant (comp. chap. 2.4). The volume fraction of species in chemical equilibrium is displayed as a function of air-/fuel ratio and temperature at constant pressure. To explain carburetion effects on emissions formation, the air-/fuel ratio distribution in the combustion chamber is determined using CFD code. This is plotted on λ-T-Maps as a function of time like can be seen in Figure 3-26 for soot volume fraction.

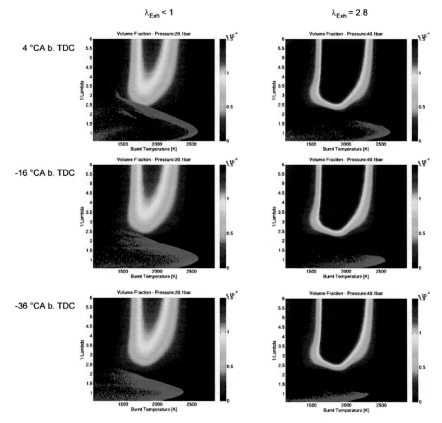

Figure 3-26: λ-T-Maps for soot volume fraction at constant pressure and CFD calculated air-/fuel ratio distribution (green dots) for engine operation under high enriched condition (λ < 1, p_{const} = 20 bar) and standard engine operation (λ = 2.8, p_{const} = 40 bar). n_{mot} = 1500 min^{-1}, imep = 4 bar, PRF 95, stratified injection

On the left side, three timesteps under high-enriched conditions are plotted. In reference to that, on the right three timesteps for the standard operation point with complete opened throttle are plotted. Like displayed in Figure 2-13, soot occures at temperatures between T = 1500 and 2300 K and at air-/fuel ratios lower than λ = 0.4 (comp. chapter 2.4). The calculated soot fraction is depending on pressure [32]. With increasing pressure, soot volume fraction increases. Under high-enriched mixture conditions (λ < 1), local conditions are reached, which result in considerable soot volume fraction resulting in a smoke number of 1.6 FSN. Comparing to that, the standard operation point (λ = 2.8) yields to quite low soot volume fraction and nearly zero smoke emissions although combustion takes place at higher pressure. Hence, with decreasing exhaust gas air-/fuel ratio, local air-/fuel ratio distribution shifts to lower air-/fuel ratio with increasing probability of soot formation. Less available oxygen reduces oxidation rate. Smoke emissions increase with decreasing air-/fuel ratio.

3.1.4 Effect of external cooled EGR on non-premixed flame temperature

Using external exhaust gas recirculation, specific heat of the cylinder mixture is increased due to the substitution of two-atomic oxygen by three-atomic components like mainly CO_2 and vaporized H_2O. The in-cylinder charge is diluted resulting in decreased oxygen content, i.e. decreased air-/fuel ratio. Hence, the absolute heat capacity of the in-cylinder mixture is increased resulting in decreased combustion temperature [31] and an increased probability of flame extinction.

Looking at Figure 3-27, using high pressure EGR at this operation point, pumping work is reduced. Exhaust back pressure is reduced because of EGR extraction in front of the turbine. Like expected, air-/fuel ratio decreases linearily with increasing EGR rate because air is substituted by exhaust gas.

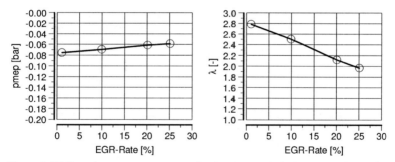

Figure 3-27: Pumping mean pressure and exhaust gas air-/fuel ratio as a function of external EGR rate. n_{mot} = 1500 min^{-1}, imep = 4 bar, Ref-O_2-03, stratified injection

Figure 3-28 shows cylinder pressure trace and mass averaged gas temperature versus crank angle degree for four external EGR rates. With increasing EGR rates pressure gradient is reduced and pressure maximum is shifted to later positions. Consequently, mass averaged temperature traces are shifted to later position. Important to notice, temperature maxima for all four EGR rates are quite similar at different timing position.

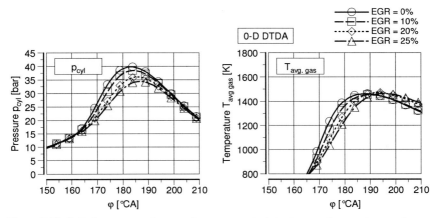

Figure 3-28: Cylinder pressure trace and mass averaged gas temperature versus crank angle degree for four external EGR rates. n_{mot} = 1500 min^{-1}, imep = 4 bar, IGA = 26.25 °CA b. TDC, Ref-O$_2$-03, stratified injection

This is explained by the corresponding decrease of air-/fuel ratio (compare Figure 3-27) as well as by the resulting rate of heat release and the resulting burn function displayed in Figure 3-29. With zero external EGR, heat release is quite quick. With increasing external EGR rate, rate of heat release is reduced resulting in extended heat release duration. Hence, with increasing external EGR rate, an increasing part of heat release takes part after top dead center resulting in an increasing mass averaged temperature in the expansion stroke.

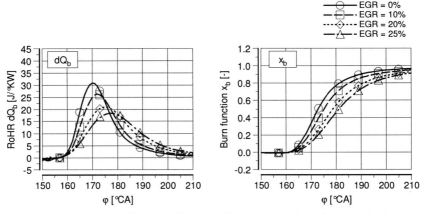

Figure 3-29: Rate of heat release and burn function versus crank angle degree for four external EGR rates. n_{mot} = 1500 min^{-1}, imep = 4 bar, IGA = 26.25 °CA b. TDC, Ref-O$_2$-03, stratified injection

This is retrieved in optical analysis of flame kernel formation and primary flame front propagation using optical fibre spark plug sensor. With increasing EGR rate, flame propagation speed $s_{f, prop}$ decreases continuously resulting in increased ignition delay and slower heat release. Like observed before and reported by Schintzel [57], with decreasing flame propagation speed propagation itself becomes radially steadier.

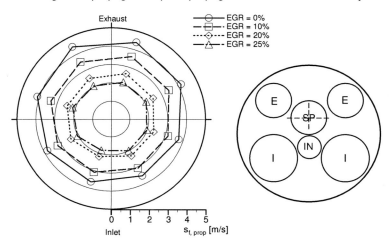

Figure 3-30: The influence of EGR rate on primary flame propagation determined using optical fibre spark plug sensor (vertical view). n_{mot} = 1500 min^{-1}, imep = 4 bar, Ref-O$_2$-03, stratified injection, Cyl.1

However, a distinctive flame kernel drift towards the exhaust can be observed approving the stable flame kernel formation independent of thermodynamic parameters.

The measured HC and CO emissions in Figure 3-31 show typical dependencies of the supplied EGR amount.

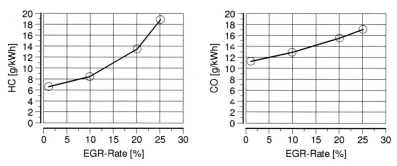

Figure 3-31: Specific HC and CO emissions as a function of external EGR rate. n_{mot} = 1500 min^{-1}, imep = 4 bar, Ref-O$_2$-03, stratified injection

Due to enhanced flame extinction probability with increasing EGR rate, HC emissions increase. The decreased combustion temperature and air-/fuel ratio reduce time and available reaction energy for back reaction of composed CO to CO_2.

These effects on engine efficiency are all well explained by a split of losses like displayed in Figure 3-32.

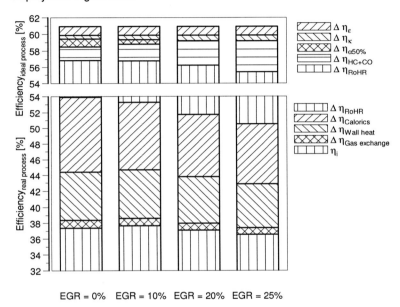

EGR = 0%   EGR = 10%   EGR = 20%   EGR = 25%

Figure 3-32: Split of losses for four external EGR rates. n_{mot} = 1500 min^{-1}, imep = 4 bar, IGA = 26.25 °CA b. TDC, Ref-O_2-03, stratified injection

With increasing EGR rate, the thermal losses due to gas composition ($\Delta \eta_\kappa$) increase due to decreasing air-/fuel ratio and increasing residual gas mass fraction. The thermal losses due to combustion phasing ($\Delta \eta_{\alpha50\%}$) are reduced since MFB50% is shifted towards top dead center (compare Figure 3-29, right diagram). The losses due to uncomplete combustion and rate of heat release ($\Delta \eta_{HC+CO}$ and $\Delta \eta_{RoHR}$) increase considerably resulting in an engine efficiency decrease towards higher EGR rates, while the calorical losses as well as the wall heat losses ($\Delta \eta_{Calorics}$ and $\Delta \eta_{Wall\ heat}$) decrease due to decreasing combustion temperature.

Hence, with increasing external cooled EGR rate, engine efficiency slightly increases at low EGR rates due to decreased pumping losses and despite increasing cyclic variations (Figure 3-33). The decrease of engine efficiency at higher EGR rates is linked to increasing losses due to higher cyclic fluctuations as well as increased heat release duration.

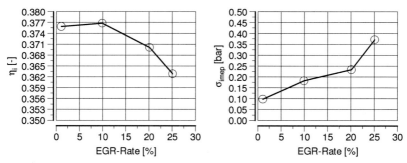

Figure 3-33: Indicated engine efficiency and standard deviation of indicated mean pressure as a function of external EGR-Rate. $n_{mot} = 1500\ min^{-1}$, imep = 4 bar, Ref-O_2-03, stratified injection

Because of the slower combustion and increasing specific heat capacity, combustion temperature decreases with increasing EGR rate. Looking at Figure 3-34, calculated temperature of non-premixed flame shows a clear tendency to decreasing flame temperature. Comparing to this, burned gas temperature shows stronger decrease of temperature $T_{burned\ gas}$. Remembering the model supposition of zero-dimensional two-zone model in chapter 2.1.3, burned gas temperature is well over-predicted due to the ideal separation of combustible air-/fuel mixture and surrounding excess air-/exhaust-/residual gas hull. With increasing EGR rate, caloric properties of the burning mixture become more similar to a one-zone diesel model with complete and spontaneous mixing of burned and surrounding gas lowering the over prediction [17]. Due to this, quite higher temperature differences for different EGR rates occur compared to T_{npf} based on thermal optical properties.

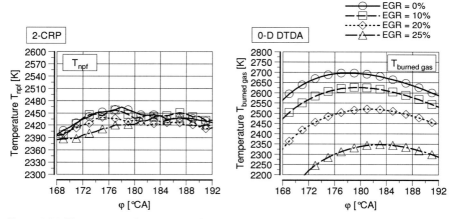

Figure 3-34: Temperature of non-premixed flame and burned gas versus crank angle degree for four external EGR rates. $n_{mot} = 1500\ min^{-1}$, imep = 4 bar, Ref-O_2-03, stratified injection, Cyl.1

Figure 3-35 shows the temperature curves of the mass mean gas temperature (one-zone model) of cylinder one for four EGR rates. Due to the mass averaged calculation including burned and unburned gas, temperature level is much lower compared to T_{npf}. With increasing EGR rate, the temperature rise is delayed. During temperature rise, a tendency of decreasing temperature with increasing EGR, similar to the tendencies found when looking at T_{npf}, can be observed.

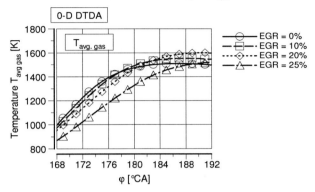

Figure 3-35: Temperature (mass averaged) of the cylinder gas versus crank angle degree for four external EGR rates. $n_{mot} = 1500 \ min^{-1}$, imep = 4 bar, Ref-$O_2$-03, stratified injection, Cyl.1

After top dead center (TDC = 180 °CA), mass averaged temperatures remain on a similar level like it can be observed for T_{npf} also. Hence, temperature decrease like observed with the optically 2-CRP determination is quite similar to the one-zone zero-dimensional mass averaged temperature. Like mentioned before, the 2-CRP determines the surface temperature of glowing soot particles. The determined non-premixed flame temperature is mainly influenced by the local temperature during formation of the particles and the on-going surface reactions. Hence, the effect of EGR on the temperature of the surrounding burned gas is much stronger than on the optically determined particle surface temperature.

NO_x emissions decrease with increasing EGR rate like displayed in Figure 3-36. Less reaction energy for NO_x composing is available and soot oxidation is diminished resulting in slightly increasing smoke emissions. By varying the amount of external EGR, a NO_x-soot trade-off can be observed emphasizing the importance of flame temperature for soot oxidation during stratified engine operation.

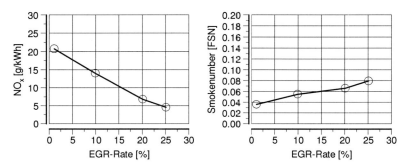

Figure 3-36: Specific NO$_x$ and smoke emissions as a function of external EGR rate. n_{mot} = 1500 min^{-1}, imep = 4 bar, Ref-O$_2$-03, stratified injection, Cyl.1

Summing up, recirculation of cooled exhaust gases greatly influences heat release and emission formation. Heat release is decelerated as well as combustion temperature is decreased. Because of mixture dilution, the absolute heat capacity of in-cylinder mixture is increased restraining NO$_x$ formation [56] and enhancing flame extinction probability. Mixture dilution as well as decreasing combustion temperature reduces soot oxidation resulting in increased soot emissions with increasing EGR rate.

3.1.5 Effect of injection strategy on soot luminosity

Using multi-injection strategy the carburetion structure and the corresponding air-/fuel ratio distribution can be influenced [12]. Therefore, application parameters are defined, which control the injected fuel mass per injection impulse. The relative fuel mass distribution between the two main injection impulses is defined as

$$\mu_1 = \frac{t_{i,1}}{t_{i,1}+t_{i,2}} \quad \text{[-] (comp. appendix b).} \qquad [3\text{-}1]$$

With increasing injection split factor μ_1 more and more fuel is shifted towards the first injection. The fraction of fuel with enlarged evaporation timescales is increased and the mixture at the spark plug is diluted. According to Jakobs [56], varying the injection splitting yields to a local air-/fuel ratio variation and resulting in changes of heat release and emission formation. To gain a maximum variation range, engine load is increased up to imep = 7 bar. The maxium reachable injection split factor, i.e. lean ignition limit, depends on engine load and speed. Engine load is directly linked to the global exhaust gas air-/fuel ratio during stratified engine operation. The maximum injection split factor decreases with decreasing engine load, i.e. increasing global exhaust gas air-/fuel ratio. Assuming a nearly constant air mass during constant engine speed and part load, a certain amount of carbureted fuel is needed at the spark plug for composing an ignitable mixture. With increasing engine speed, the effect of increasing turbulence becomes dominant limiting the range of possible injection split factors [12].

Figure 3-37 shows the indicated engine efficiency and the standard deviation of indicated mean pressure as a function of the injection split factor, i.e. qualitative local air-/fuel ratio. With increasing injection split factor, engine efficiency increases considerably. Beyond an injection time ratio of 0.7, engine efficiency remains constant, while the standard deviation increases again. The lean ignition limit is reached. A further increase of the injection split factor results in misfire occurrence.

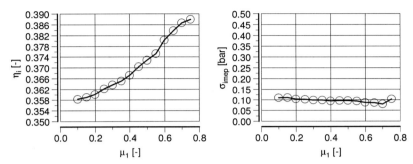

Figure 3-37: Indicated engine efficiency and standard deviation of indicated mean pressure as a function of injection split factor. $n_{mot} = 1500\ min^{-1}$, imep = 7 bar, Ref-O_2-03, λ = 1.7, stratified injection

Figure 3-38 displays the calculated ignition delay and the maximum pressure gradient as a function of injection split factor while Figure 3-39 shows cylinder pressure trace and mass averaged gas temperature versus crank angle degree for four injection split factors. A marginal tendency to increased ignition delay time can be observed, while the maximum pressure gradient decreases distinctively at $\mu_1 >$ 0.4.

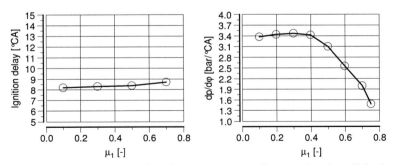

Figure 3-38: Ignition delay and maximum pressure gradient as a function of injection split factor. $n_{mot} = 1500\ min^{-1}$, imep = 7 bar, Ref-O_2-03, λ = 1.7, stratified injection

Corresponding to that, with increasing injection split factor, i.e. shifting the required fuel mass from the second to the first injection impulse, pressure trace remains similar up to an injection split factor of 0.3. With further increase of the injection split factor, pressure trace is shifted to later positions resulting in decreased mass averaged gas temperature.

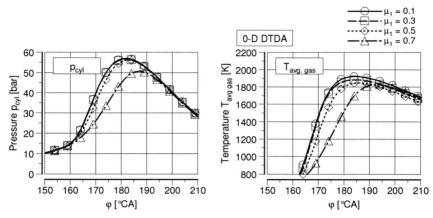

Figure 3-39: Cylinder pressure trace and mass averaged gas temperature versus crank angle degree for four injection split factors. n_{mot} = 1500 min^{-1}, imep = 7 bar, IGA = 28.875 °CA b. TDC, Ref-O_2-03, λ = 1.7, stratified injection

Looking at the resulting rates of heat release and burn functions for four injection split factors in Figure 3-40, strong influences of the modified carburetion structure can be found. With increasing injection split factor heat release decelerates continuously. Consequently, combustion is shifted to later positions and lasts considerably longer.

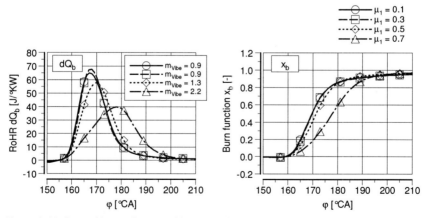

Figure 3-40: Rate of heat release and burn function versus crank angle degree for four injection split factors. n_{mot} = 1500 min^{-1}, imep = 7 bar, IGA = 28.875 °CA b. TDC, Ref-O_2-03, λ = 1.7, stratified injection

The measured HC and CO emissions in Figure 3-41 decrease up to μ_1 = 0.4. Beyond this, HC emissions increase exponentially while CO emissions decrease up to μ_1 = 0.6. This can be explained looking at ignition delay, pressure gradient, pressure traces and the resulting rate of heat releases in Figure 3-38, Figure 3-39 and Figure

3-40. Due to the deceleration of heat release beyond $\mu_1 = 0.4$, the probability of flame quenching increases. Decreasing CO emissions indicate increasing combustion efficiency and increasing local air-/fuel ratio.

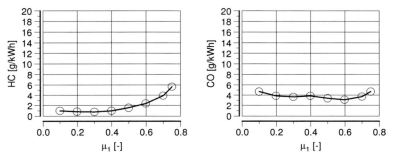

Figure 3-41: Specific HC and CO emissions as a function of injection split factor. $n_{mot} = 1500$ min^{-1}, imep = 7 bar, Ref-O_2-03, $\lambda = 1.7$, stratified injection

Hence, engine efficiency increase is caused by two effects. At injection split factors $\mu_1 < 0.4$, increased combustion efficiency increase engine efficiency expressed by the decreased HC and CO emissions (compare Figure 3-44). At injection split factors $\mu_1 > 0.4$, engine efficiency increase is caused by later combustion phasing as well as reduced mass averaged gas temperature reducing caloric and wall heat losses.

Next to the deceleration of heat release, the shape of the rate of heat release changes with enleaning local spark plug air-/fuel ratio. At low injection splitting factor, i.e. quite enriched mixture at the spark plug, heat release curve shows a characteristical shape for stratified engine operation. Due to the strong gradient of air-/fuel ratio structure with an enriched mixture around the spark plug, a high rate of heat release during the first period of combustion can be observed. The second period is characterized by a burnout at low rates of heat release [54]. With increasing injection split factor, the shape of heat release becomes more and more symmetrical at lower rates of heat release. Using the Vibe forming factor m [174] to describe the symmetry of heat release over time, it becomes evident that with increasing injection splitting factor the shape of heat release becomes similar to the shape of heat release of an SI-engine with homogenous mixture (comp. appendix d).

Figure 3-42 shows the optical determined primary flame front as a polar diagram of the flame propagation speed $s_{f, prop}$. Like observed before, the formed flame kernel drifts towards the exhaust side. With increasing injection split factor, flame propagation remains similar, although heat release is decelerated. In opposition to the influences of fuel pressure, exhaust gas air-/fuel ratio or EGR, flame propagation speed does not correlate with rate of heat release. Though homogenization of the mixture is changed with changing the injection split factor, inflammation conditions at the spark seems to remain constant due to the set post injection into the inflaming mixture.

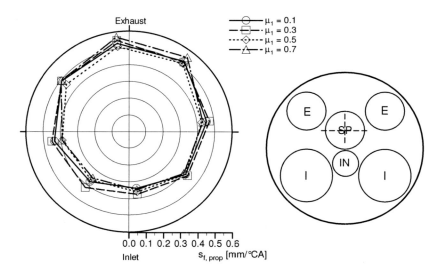

Figure 3-42: The influence of injection split factor on primary flame propagation determined using optical fibre spark plug sensor (vertical view). $n_{mot} = 1500\ min^{-1}$, imep = 7 bar, Ref-O_2-03, $\lambda = 1.7$, stratified injection, Cyl.1

While fuel pressure, exhaust gas air-/fuel ratio and EGR seem to influence imflammation superiorily, the degree of mixture homogenization does not influence the primary flame propagation. This is approved by the nearly constant ignition delay during variation of the injection split factor (compare Figure 3-38).

Conjoined with the observations of Jakobs [56] regarding the change of air-/fuel ratio structure due to injection strategies, it is assumed that with increasing injection split factor the local air-/fuel ratio at the spark plug is changed and the change of fuel mass distribution on the main injection impulses results in a more homogenous mixture. Thus, the shape of heat release becomes more symmetrical. This can be retrieved in the optically detected soot luminosity fraction using 2-CRP in Figure 3-43. At low injection split factors, a strong luminosity signal is detected. With increasing injection split factor, the level of detectable soot luminosity decreases. At the lean ignition limit with $\mu_1 = 0.7$, soot luminosity reaches the detection limit. The occurrence of non-premixed combustion phases is nearly zero, indicating a quite homogenous mixture with a lean local air-/fuel ratio.

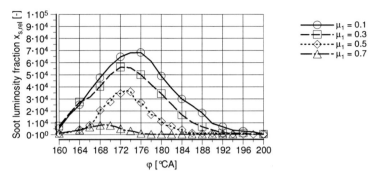

Figure 3-43: Relative soot luminosity fraction versus crank angle degree for four injection split factors. n_{mot} = 1500 min^{-1}, imep = 7 bar, Ref-O$_2$-03, λ = 1.7, stratified injection , Cyl.1

This is approved by the deveolpings of the raw emissions when increasing injection split factor in Figure 3-44. NO$_x$ emissions slightly increase. The local air-/fuel ratio shifts slightly from enriched condition towards leaner condition resulting in increasing NO$_x$ emissions. Beyond μ_1 = 0.5, NO$_x$ emissions decrease strongly. The local air-/fuel ratio for maximum NO$_x$ emissions ($\lambda \approx$ 1.1) is passed. Additionally, due to the decelerated heat release, combustion phasing is shifted to later positions (compare Figure 3-40) leading to further decreae of NO$_x$ emissions. Next to this, soot emissions decrease considerably. At μ_1 = 0.7, no smoke emissions are detectable congruent to the optically detected soot luminosity fraction. CO emissions decrease up to an injection split factor of μ_1 = 0.6 indicating a leaner and more homogenous local mixture. At $\mu_1 >$ 0.6, CO emissions increase again indicating a local air-/fuel ratio $\lambda >$ 1.2 (compare Figure 2-10). The mixture is more homogenous and leaner but also broader distributed. Hence, less soot is formed, but the flame front comprehends less of the mixture resulting in the discussed increased HC emissions.

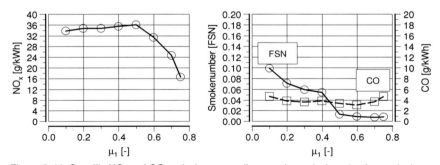

Figure 3-44: Specific NO$_x$ and CO emissions as well as smoke emissions (entire engine) as a function of injection split factor. n_{mot} = 1500 min^{-1}, imep = 7 bar, Ref-O$_2$-03, λ = 1.7, stratified injection

Using optimized multiple injection strategies, shape of heat release can be modeled in certain ranges. Aspiring to the maximum possible homogenization of the stratified mixture, NO$_x$ and soot emissions can be decreased and engine efficiency increased simultaneously. Furthermore, periods of non-premixed combustion are suppressed and avoid the formation of soot. The homogenization is limited by a certain amount of fuel at the spark plug. This is needed for creating a sufficient ignitable mixture to initiate the combustion. Hence, with increasing engine load, the degree of homogenization can be increased. Increasing HC and CO emissions point to the lean ignition limit.

Summary of engine parameters effects

Analysis of the primary flame propagation shows a faster propagation towards the exhaust. With closed tumble flap, a slight drift and increased flame propagation speed can be observed. Decreasing exhaust gas air-/fuel ratio and fuel pressure results in a decreased flame propagation speed. Flame propagation becomes steadier during flame kernel formation. In opposition to that, an increased degree of homogenization caused by changing injection timings does not effect the primary flame propagation. Increasing EGR rate yields to continously decreasing flame propagation speed.

Closing the tumble flap and changing the fuel pressure result in marginal changes of non-premixed flame temperature caused by changing heat release barycenters. A decrease of exhaust gas air-/fuel ratio and an increase of EGR rate result in decreasing non-premixed flame temperature. Comparing to the zero-dimensional burned gas temperature, temperature decrease is lower due to the model assumption of the used two-zone-model.

Relative soot luminosity fraction increases with decreasing exhaust gas air-/fuel ratio and fuel pressure in agreement with the measured smoke emissions. By shifting relative fuel mass from the second to the first injection, homogenization is improved and local air-/fuel ratio is increased. Relative soot luminosity fraction decreases continuously and smoke emissions decrease until detection limit is reached. The shape of heat release becomes symmetrical and reaches Vibe forming factors comparable to homogenous engine operation.

Regarding the influence of engine operating parameters on the level of non-premixed flame temperature as well as its distribution, no influence of engine operating parameters like air-/fuel ratio, rail pressure or external cooled EGR on the shape of non-premixed flame temperature distribution ρ_{Tnpf} can be observed. Thus, although the temperature of the distribution's maximum slightly changes depending on the thermodynamic conditions, the shown non-premixed flame temperature distribution is characteristical for non-premixed combustion of spray-guided stratified mixture. Optically determined non-premixed flame temperature is driven by the thermal behavior of the formed soot particles and its surface reactions due to contemporaneous soot formation and oxidation.

3.1.6 Carburetion study of ethanol enriched dual and multi component fuels for stratified mixture formation

Fuel composition greatly influences combustion characteristics. Like described in chapter 2.1.1, a reduction of C/H ratio and an increase of fuel bounded oxygen increases the specific heat of the resulting mixture. This results in lower combustion gas temperature, less dissociation losses and influences composing and oxidation of temperature driven emission formation like CO and NO_x.

Concerning soot production and oxidation in diesel engines, several fuel studies are published [80] referring to oxygen enrichment of fuel like DME and other oxygenating components [113],[116]. All authors report a remarkable reduction of soot emissions. In gasoline engines ethanol components are widely used as oxygenating additive. Ethanol can be produced using alternative refinery processes and is expected to be conducive to decrease fossil based carbon dioxide emissions [175]. Thermodynamically it is characterized by an increased evaporation enthalpy, an increasing specific heat in the resulting burned gas and a low heating value. Due to the lower heating value an increased fuel mass is needed for equal energy supply leading to a further increased energy demand for fuel vaporization and increasing absolute heat. In direct injecting SI-engines this energy has to be detracted from the cylinder mixture, i.e. its internal energy. The mixture is cooled down resulting in lower compression and combustion gas temperature. Under full-load condition, a decreased knock tendency is reported due to this effect [176]. During engine start, especially under engine cold conditions, there is not enough internal energy available resulting in strong carburetion artifacts and misfire occurring [176],[180]. Aleiferis et al. [181] have performed optical studies of carburetion and combustion using dual-component fuels in a DISI engine during homogenous engine operation. Some publications deal with the influence of commercial E85 fuel on stratified engine operation. Bäcker et al. [175] have performed optical carburetion and combustion studies on a single-cylinder engine with a spray-guided combustion process. They found a decrease of NO_x emissions of about fourty percent during stratified engine operation and an increased lean and EGR tolerance. Sandquist et al. have [182] investigated the influence of low ethanol fraction on stratified engine operation using an entire engine with a wall-/air-guided combustion process. They report about zero smoke emissions and increased HC emissions when adding ethanol. Smith et al. [183] have compared ethanol with iso-buthanol and iso-octane in an optical single-cylinder engine with a spray-guided combustion process. They report of a slightly smaller stable operating range when using mono-component alcohol fuels. Synthetic fuel design offers further possibilities to increase combustion efficiency and decrease pollutant emissions. Thus, next to the influence of ethanol on dual-component fuels, a synthetic multi-component fuel should be considered [184].

With regard to stratified engine operation and its demand for a quick and local stable fuel evaporation and carburetion, ethanol injection is expected to influence stratified carburetion [175],[183]. To describe evaporation and carburetion characteristics and evaluate their feasible influence on engine operation, a fuel study in a high-pressure injection chamber (see appendix j) is accomplished. Table E-7-5 (see appendix e) displays the used fuel compositions. As multi-component reference, the oec-reference fuel (Ref-O_2-03) is used. It is characterized by low oxygen content (<0.3 vol. %), a research octane number (RON) of 100 and an aromatic hydrocarbon content of thirty percent. To exclude interacting influences of different hydrocarbon compounds on evaporation, carburetion, combustion and emission formation iso-octane (i-C8) as 2,2,4-trimethylpentane is used. By adding ethanol in two stages to iso-octane (twenty and eighty-five percent ethanol) ethanol influence on internal engine processes can be described. Finally, a synthetic multi-component (DIGAFA) fuel is used. It is characterized by a RON of 100, an increased oxygen content and low aromatic hydrocarbon content.

Using a combined Mie/Schlieren measurement technique (comp. appendix i), liquid and gaseous fuel phases can be visualized qualitatively and their penetration can be determined under thermodynamically constant conditions. Figure 3-45 shows an image sequence of Mie/Schlieren visualization during high-pressure camber fuel injection of reference fuel (Ref-O_2-03), iso-octane (i-C8), twenty percent ethanol added (E20/i-C8), eighty-five percent ethanol added (E85/i-C8) and synthetic gasoline (DIGAFA). The camera system is aligned to the injector axis with the injector nozzle in the center of the view. As green circular background a mirror can be recognized. The image sequence shows records for four sequent time steps. After a short initialization delay liquid fuel is detected penetrating circularry into the pressure chamber. During that already a slight green corona can be observed displaying gaseous evaporated fuel. At the third time step obvious differences can be recognized. During Ref-O_2-03 injection, only a small gaseous circle is visualized. In comparison to that, with i-C8, a strong green corona with large content of evaporated fuel occurs. Additionally, a shorter penetration of the liquid phase can be seen. With increasing ethanol content the liquid penetration increases and the detectable evaporated fuel content decreases. The gaseous penetration does not differ that much. The last time step is recorded after end of injection. The remaining liquid fuel forms a quader due to the injector design. Around the liquid phase as well as inside evaporation takes place. Remarkably, with i-C8, almost all the fuel is evaporated. With increasing ethanol content, the content of remaining liquid fuel increases, i.e. less progress of the evaporation process can be observed. The Ref-O_2-03 fuel shows a similar progress to E20/i-C8.

Consequently, the highest evaporation rate can be observed during i-C8 injection. Increasing ethanol content decreases the evaporation rate due to its higher evaporation enthalpy. The multi-component Ref-O_2-03 fuel shows similar evaporation behaviour to E20/i-C8. It should be pointed out, that all fuel injections result in quite similar penetrations of the gaseous phase. Hence, no modification of spark-plug position is necessary for different fuel use. Only different air-/fuel distribution can be expected and thus necessitate smart changes of injection strategy. Referring to changes in injection masses due to different heating values of the different fuels, changing injection times needs to be considered. With decreasing heating value of the fuel, more fuel mass is needed and the injection time increases. Looking at Figure 3-46 (a), the radial penetration is not influenced by the injection time. Hence, the correction of the changing heating value is not expected to greatly influence engine operation. In opposition to the injection time, chamber temperature (b), i.e. cylinder temperature as well as the fuel rail pressure (c) influences the radial penetration of the liquid fuel. With increasing medium temperature, radial penetration decreases. Hence, under certain engine operation conditions, a mixture heating by uncooled EGR is able to improve evaporation and carburetion processes. An increase of the fuel pressure results in a slight decrease of liquid fuel penetration. Like discussed before, increasing fuel pressure lets expect smaller droplets and therewith improved carburetion. The streaming impuls is lowered because of the lower droplet mass. The slight earlier decrease of detected liquid fuel indicates an equal assumption. This also verifies the improvement of carburetion, which can be observed in chapter 3.1.2 discussing the effect of fuel rail pressure on engine performance. Referring to the influence of the needle lift of the injector (d), an increasing maximal radial penetration can be found with increasing needle lift, i.e. decreasing duty cycle (DC). With increasing needle lift, the atomization of the liquid fuel is reduced. Larger droplets with a higher mass are generated implicating a higher mass impulse when penetrating into the chamber. Thus, a higher penetration can be observed very similar to the influence of decreasing medium temperature.

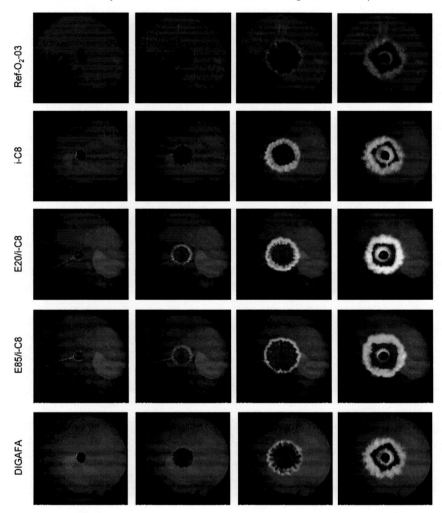

Figure 3-45: Image sequence of Mie/Schlieren visualization during high-pressure chamber fuel injection of Ref-O$_2$-03, i-C8, E20/i-C8, E85/i-C8 and DIGAFA. T_{chamb} = 600 K, p_{chamb} = 10 bar, p_{Rail} = 200 bar, t_i = 0.45 ms

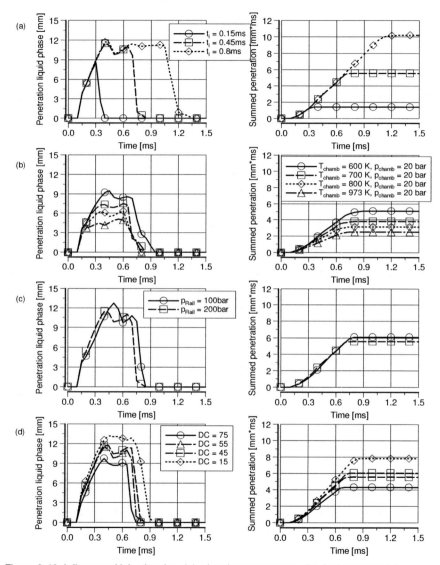

Figure 3-46: Influence of injection time (a), chamber temperature (b), fuel pressure (c), needle lift (DC) (d) on radial penetration of liquid fuel under compression injection condition. *i-C8, Basic settings: t_i = 0.45 ms, T_{chamb} = 600K, p_{chamb} = 10 bar, p_{Rail} = 200 bar, DC = 55.*

Summing up, it should be pointed out, that an increasing ethanol fraction results in decelerated evaporation. For heating value correction the injection time needs to be adjusted. An adjustment of the fuel flow via adjustment of the needle lift results in increasing radial penetration and in undesirable changes of mixture structure and composition. Changes of the fuel pressure only shows slight tendencies in regard to the liquid radial penetration. The increasing vapor enthalpy with increasing ethanol fraction also gives an increasing cooling effect on the cyclinder mixture. The energy needed for evaporation is taken from the internal energy of the cylinder mixture resulting in a temperature decrease. Hence, combustion starts at a lower temperature level resulting in decreased combustion temperature. This is enhanced even more by the increased specific heat with increasing ethanol fraction (compare chapter 2.1.1). Regarding emission formation, decreased nitric oxide and increased carbon monoxide emissions are expected. The second characteristic of ethanol enriched fuel is the fuel bounded hydroxyl group. Due to this, a distinctive enhancement of soot oxidation is expected (comp. chapter 2.4).

3.1.7 The effect of ethanol addition on non-premixed flame temperature

Regarding to the influence of fuel modification on optical combustion characteristic, one major effect is the decrease of combustion luminosity in the visible and near infrared spectra. While engine operation with the multi-component reference fuel results in strong luminosity of composed soot particles, the addition of ethanol results soot luminosity. When using E85/i-C8, no combustion luminosity in the chosen wavelength range can be detected even at enriched mixture conditions.

Figure 3-47 shows calculated non-premixed flame temperature distribution of single records during stratified injection engine operation using the multi-component reference fuel (Ref-O_2-03), iso-octane (i-C8) and ethanol enriched dual component fuel (E20/i-C8). Quite characteristical, the images of Ref-O_2-03 show the known temperature distribution. Typical flame structures of non-premixed flames can be recognized with explicit flame streams. Towards the edges of zones of non-premixed flame, temperature increases like discussed in chapter 2.6. Looking at the image sequence of i-C8, the temperature distribution is quite similar like discussed in chapter 2.6. In comparison to Ref-O_2-03, the flame structure cannot be recognized that clear. The illustratable soot luminosity is much more jagged giving the impression of less generated soot, which is expectable due to the fuel composition. More differences can be found in the image sequence of the calculated non-premixed flame temperature of E20/i-C8. The images are much more diffuse. This is caused by the decreased combustion luminosity in the visuable wavelength range. To record evaluable light information, the gate of the image intensifier needs to be increased by a factor of one hundred in comparison to the records of Ref-O_2-03. Thus, the images are taken over a timescale of more than one crank angle degree. Consequently, the resulting luminosity distribution and calculated temperature distribution does not show clear flame structures. The images show temperature distributions similar to the discussed optical characteristics with a cold zone beneath the injector and a hotter zone towards the inlet, but the calculated flame temperature is lower in comparison to the reference fuel and to iso-octane.

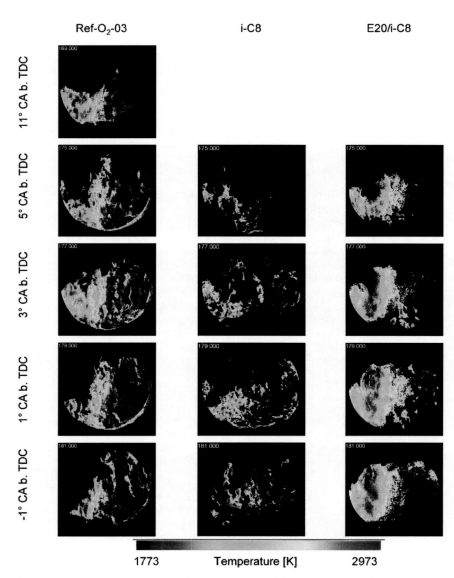

Figure 3-47: Image sequence of calculated non-premixed flame temperature T_{npf} distribution using multi-component reference fuel (Ref-O_2-03), iso-octane (i-C8) and dual-component fuel (E20/i-C8). $n_{mot} = 1500\ min^{-1}$, imep = 4 bar, $\lambda = 2.8$, stratified injection, Cyl.1; For E20/i-C8: increased gate of the image intensifier of factor one hundred

Calculating the image averaged non-premixed temperature as a function of crank angle degree, a temperature offset of about one hundred Kelvin between i-C8 and E20/i-C8 can be found like displayed in Figure 3-48. The non-premixed flame temperature of Ref-O_2-03 shows a maximum around top dead center. The maximum of i-C8 and E20/i-C8 can be found between five and ten crank angle degrees after top dead center.

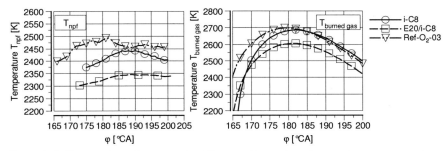

Figure 3-48: Temperature of non-premixed flame and burned gas temperature versus crank angle degree for three fuels. n_{mot} = 1500 min^{-1}, imep = 4 bar, λ = 2.8, stratified injection, Cyl.1

Looking at the calculated zero-dimensional temperature of the burned zone, the same trends can be retrieved. Corresponding to the non-premixed flame temperature calculation, E20/i-C8 yields to a comparable temperature decrease in comparison to i-C8, while Ref-O_2-03 results in an earlier temperature maximum with a maximum temperature similar to i-C8. Like observed before, the optical measurements based non-premixed flame temperature is about two hundred to two-hundred fifty Kelvin lower than the zero-dimensional calculated burned gas temperature.

The qualitative impression of less detectable soot luminosity is verified by the curves of soot luminosity fraction in Figure 3-49. Corresponding to the temperature curves, an early maximum of the multi-component reference fuel at about four crank angle degrees before top dead center can be observed, while the maximum of iso-octane is delayed. Quite characteristic, the time of maximum soot luminosity is a bit earlier than the time of temperature maximum. The soot luminosity fraction measured during Ref-O_2-03 fuel operation is nearly twice as high as with i-C8 operation. Because of the dramatically increased gate during E20/i-C8 operation, no comparable evaluation can be performed. The other way around, using comparable record timescales, no soot luminosity can be detected when using E20/i-C8.

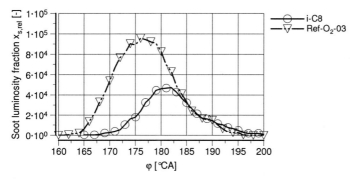

Figure 3-49: Relative soot luminosity fraction versus crank angle degree for multi-component reference fuel (Ref-O_2-03) and iso-octane (i-C8). n_{mot} = 1500 min^{-1}, imep = 4 bar, λ = 2.8, stratified injection, Cyl.1

This becomes more evident when looking at the detected combustion light intensity using the optical fibre spark plug sensor in Figure 3-50. Maximum intensity is twice as high when using Ref-O_2-03 compared to DIGAFA. Operating the engine with i-C8, the averaged intensity reaches just seventeen percent of the maximum intensity. Ethanol addition leads to quite low intensity, which is mainly generated by chemiluminiscence of gaseous radicals.

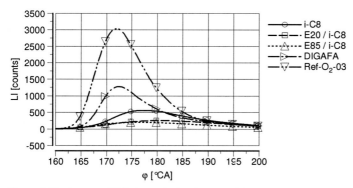

Figure 3-50: Averaged light intensity for five tested fuels detected by the optical fibre spark plug sensor. n_{mot} = 1500 min^{-1}, imep = 4 bar, λ = 2.8, stratified injection, Cyl.1

Summing up, the addition of ethanol decreases combustion temperature like expected caused by its increased evaporation enthalpy and specific heat. Referring to optical characteristics, the luminosity distribution in the detected view stays similar approving the dominant influence of the spray characteristic on combustion, while fuel modification yields to changes of soot luminosity. Eliminating aromatic content in the fuel by the use of i-C8, a jagged flame structure can be observed. The addition of ethanol decreases soot luminosity excessively. The detectable light intensity is of about one hundred times lower. Reminding the decreased evaporation tendency of ethanol enriched fuel, a similar content of non-premixed mixture can be expected.

Furthermore, the decreased combustion temperature results in slightly reduced flame luminosity. During EGR variation, a luminosity factor of two can be observed. Hence, the excessively decreased soot luminosity can only be reasoned by considerably reduced soot formation and enhanced soot oxidation.

3.1.8 The influence of ethanol enriched dual- and multi-component fuels on stratified combustion

The observed influence of fuel modification on soot luminosity and non-premixed flame temperature corresponds to further characteristics of heat release, emission formation and the resulting engine efficiency. Starting with opened throttle at standard engine operation point, the throttle is closed continuously at constant engine load for varying the exhaust gas air-/fuel ratio. The air mass flow is decreased linearily. Expressing the air mass flow using a linear equation with the linear coefficients a and b, the fuel mass flow can be described using equation 2-15 as

$$m_F = \frac{a \cdot \lambda + b}{\lambda \cdot L_{st}}$$

[3-2]

with a,b linear coefficients depending on engine operation point.

Figure 3-51 shows the calculated fuel mass flow and its derivations as a function of exhaust gas air-/fuel ratio. Linear coefficients are determined using a linear regression of the measured air mass flow. With increasing L_{st}, the negative gradient of fuel mass flow decreases. At air-/fuel ratios smaller than $\lambda \approx 2$, the negative gradient increases strongly. Setting the characteristical gradient threshold to $dm_F/d\lambda$ = -0.5, a characteristical exhaust gas air-/fuel ratio of around $\lambda \approx 1.6$ can be found depending on the fuel. Below $\lambda \approx 1.6$, fuel mass flow increases distinctively.

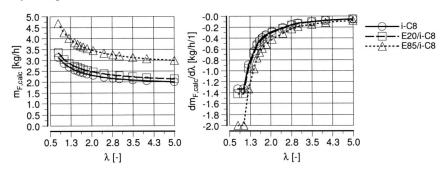

Figure 3-51: Calculated mass fuel flow and their derivations for the mono-/dual-component fuels.

Figure 3-52 shows the indicated engine efficiency during stratified engine operation with the dual-component fuels (left) and the multi-component fuels (right) as a function of exhaust gas air-/fuel ratio. Engine efficiency decreases with decreasing exhaust gas air-/fuel ratio. Up to $\lambda \approx 1.6$, a quite linear decrease can be observed. Below $\lambda \approx 1.6$, fuel mass flow increases exponentially corresponding to the predicted fuel mass flow. Looking at the dual component fuels, engine efficiency decreases with increasing ethanol fraction. At open throttle, engine efficiency of more than 0.38

can be reached. Using the multi component fuels, only slight differences can be observed. Engine efficiency is higher than with i-C8 and E20/i-C8 but lower than using E85/i-C8.

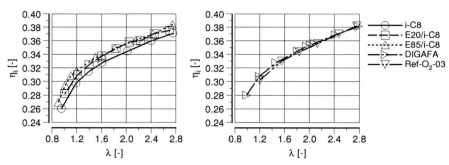

Figure 3-52: Indicated engine efficiency as a function of exhaust gas air-/fuel ratio. n_{mot} = 1500 min^{-1}, imep = 4 bar, stratified injection

Figure 3-53 shows the specific CO_2 emissions during exhaust gas air-/fuel ratio variation. Towards low exhaust gas air-/fuel ratios, an increasing ethanol fraction results in increasing CO_2 emissions. At higher air-/fuel ratios the differences diminish. At understoichiometrical conditions, CO_2 emissions decrease again when using i-C8. Using the multi component fuels, the reference fuel (Ref-O$_2$-03) shows a constant offset towards higher CO_2 emissions in comparison to the synthetic fuel (DIGAFA). At standard operation condition, the CO_2 emissions of the multi-component fuels are higher than the CO_2 emissions of the dual component fuels. At rich condition, E85/i-C8 shows the highest CO_2 emissions. This has to be seen with the corresponding CO emissions.

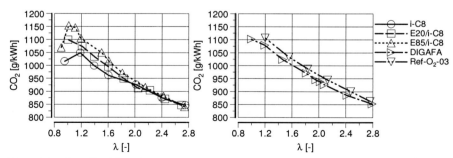

Figure 3-53: Specific CO_2 emisions as a function of exhast air-/fuel ratio. n_{mot} = 1500 min^{-1}, imep = 4 bar, stratified injection

Figure 3-54 shows the CO emissions as a function of exhaust air-/fuel ratio. In the left diagram, a strong peak at enriched conditions when using iso-octane (i-C8) can be found explaining the characteristic decrease of CO_2 emissions at this operation point. CO emissions are observed to be minimal for the characteristic exhaust gas air/fuel ratio of $\lambda \approx 1.6$. With increasing ethanol fraction, the CO emissions decrease at $\lambda < 2$.

CO oxidation is enhanced via the water gas shift reaction. Comparing the CO emissions when using the synthetic fuel, nearly equal CO emissions can be noticed. The synthetic fuel shows a quite similar oxygen fraction approving the enhancing effect of fuel bounded oxygen on CO oxidation. Towards higher air-/fuel ratios, only slight tendencies can be observed. Looking at the dual-component fuel, slightly increased CO emissions can be identified with increasing ethanol fraction. Due to the increased specific heat of the fuel and the mixture cooling during evaporation, combustion temperature is lowered. Enhanced flame quenching can be assumed. Depending on the exhaust gas air-/fuel ratio, the influences of the water gas shift reaction or flame quenching effect the CO emissions and their dependence on ethanol fraction.

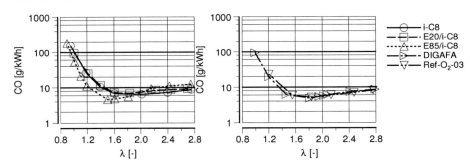

Figure 3-54: Specific CO emissions in logarithmic scale as a function of exhaust gas air-/fuel ratio. n_{mot} = 1500 min^{-1}, imep = 4 bar, stratified injection

Next to CO emissions, NO_x emissions are quite sensitive to oxygen fraction and combustion temperature. Figure 3-55 shows the NO_x emissions as a function of exhaust air-/fuel ratio. At standard engine operation point (λ = 2.8), NO_x emissions decrease with increasing ethanol content. With E20/i-C8 just a slight tendency can be observed due to an earlier combustion time (compare Figure 3-57). With E85/i-C8, distinctively decreased NO_x emissions are found. Comparing to the multi-component fuels, NO_x emissions when using Ref-O_2-03 and DIGAFA are on a similar level like i-C8. Differences of NO_x emissions are reasoned by slightly different combustion times.

With decreasing exhaust gas air-/fuel ratio, NO_x emissions develop with a distinctive maximum. The mean combustion air-/fuel ratio is decreased traversing the characteristical NO_x emissions curve with a maximum at a mean combustion air-/fuel ratio of about $\lambda \approx 1.1$ (compare Figure 2-10). I-C8 shows the highest NO_x emissions at maximum. With increasing ethanol fraction, the maximum of NO_x emissions is decreased. Using the multi-component fuels, the maxima can be found at an exhaust gas air-/fuel ratio around $\lambda \approx 2$. With i-C8, a maximum around λ = 2 can be observed. With increasing ethanol fraction this maximum is shifted to lower air-/fuel ratios resulting in a maximum of $\lambda \approx 1.6$ for E85/i-C8.

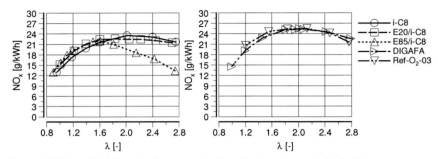

Figure 3-55: Specific NO$_x$ emissions as a function of exhaust gas air-/fuel ratio. n_{mot} = 1500 min^{-1}, imep = 4 bar, stratified injection

This is reasoned by the increased exhaust gas mass flow when using enriched fuels due to the decreasing lower heating value. Looking at the measured NO$_x$ concentrations and their derivations in Figure 3-56, all fuels show a maximum of NO$_x$ concentrations at the characteristical λ ≈ 1.6.

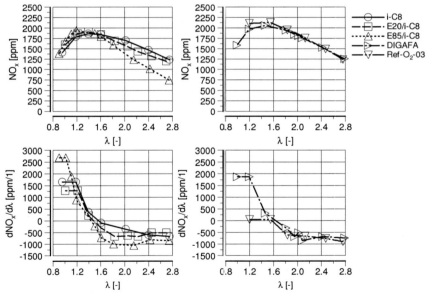

Figure 3-56: Measured NO$_x$ concentrations as a function of exhaust gas air-/fuel ratio and their derivations. n_{mot} = 1500 min^{-1}, imep = 4 bar, stratified injection

With increasing ethanol fraction, the maximum is slightly shifted to lower exhaust gas air-/fuel ratios. At maximum NO$_x$ concentrations, these are higher with increasing ethanol fraction. This is reasoned by earlier combustion time due to an accelerated heat release like can be observed in Figure 3-57. With decreasing exhaust gas air-

/fuel ratio, heat release is decelerated resulting in later combustion times except during engine operation with E85/i-C8. Until $\lambda \approx 1.6$, MFB 50% remains constant using E85/i-C8 explaining the narrowing of the NO_x emissions curves of i-C8, E20/i-C8 and E85/i-C8. Below $\lambda \approx 1.6$, heat release is decelerated using E85/i-C8 synchroneously to i-C8 and E20/i-C8. Still heat relase is faster explaining the slightly higher NO_x emissions at $\lambda < 1.6$.

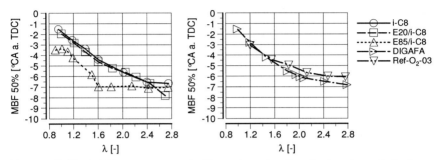

Figure 3-57: Time of fifty percent mass fuel burned as a function of exhaust gas air-/fuel ratio. $n_{mot} = 1500\ min^{-1}$, imep = 4 bar, stratified injection

Next to the decreased NO_x emissions due to increased specific heat, ethanol addition shows the expected improved soot oxidation. During optical analysis of E85/i-C8 no phasis of non-premixed combustion are detectable. Looking at Figure 3-58 displaying the measured smoke emissions as a function of exhaust gas air-/fuel ratio, this becomes more evident. Even at stoichiometrical conditions during stratified injection, no smoke emissions can be detected when using E85/i-C8.

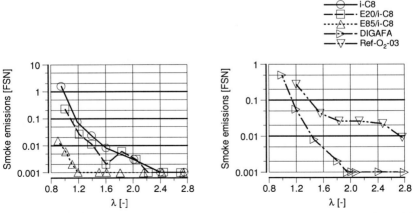

Figure 3-58: Smoke emissions as a function of exhaust gas air-/fuel ratio. $n_{mot} = 1500\ min^{-1}$, imep = 4 bar, stratified injection

Runnning with i-C8 results in increased smoke emissions under this mixture condition even compared to the multi-component fuels. Important to notice, that smoke emissions start to increase at air-/fuel ratios smaller than 1.6. DIGAFA shows a developing quite equal to E20/i-C8 which is characterized by a similar C/H/O ratio. Hence, a qualitative correlation between oxygen content and smoke emission tendency can be found.

Summing up, ethanol addition results in increased engine efficiency with decreased NO_x emissions. Even under enriched conditions, no smoke emissions are detectable. Depending on the exhaust gas air-/fuel ratio, the influences of the water gas shift reaction or flame quenching effect the CO and CO_2 emissions and their dependence on ethanol fraction. Considering multi-component fuels, the synthetic fuel (DIGAFA) shows slightly increased engine efficiency as well as decreased NO_x emissions in combination with decreased CO_2 emissions and smokeless combustion down to an exhaust gas air-/fuel ratio of $\lambda > 1.5$.

3.1.9 The inflammability limits of ethanol enriched dual- and multi component fuels

Next to energy efficiency and emission formation tendency, inflammability and the consequential cyclic variations are quite decisive for engine efficiency under stratified operation mode. Due to overstoichiometrical engine run, engine internal NO_x reduction has to be investigated. Using external cooled EGR, the specific and absolute heat of the cylinder mixture is increased and air-/fuel ratio is decreased due to the substitution of air by exhaust gas like displayed in Figure 3-59.

The resulting combustion temperature is lowered resulting in decreased NO_x emissions. EGR is mostly inert and increases the probability of flame extinction and misfire occurrence. Hence, the achievable EGR rate is an important evaluation parameter for stratified combustion.

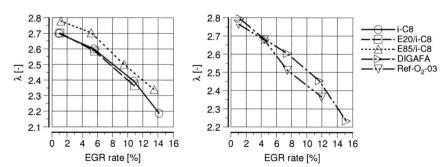

Figure 3-59: Exhaust gas air-/fuel ratio as a function of external EGR rate. $n_{mot} = 1500\ min^{-1}$, imep = 4 bar, stratified injection

Figure 3-60 shows the indicated engine efficiency for an EGR variation using three dual-component and two multi-component fuels. Important to notice, that the use of E20/i-C8 and reference fuel (Ref-O_2-03) results in a maximum EGR rate of eleven to twelve percent, while the use of DIGAFA allows a maximum EGR rate of fifteen percent. Even more, DIGAFA results in an engine efficiency of about 0.385 very similar to E85/i-C8. Iso-Octane (i-C8) results in the lowest engine efficiency. All fuels except E85/i-C8 show a quite characteristic developing of the engine efficiency versus EGR rate. At first, a slight increase can be noticed. Pumping losses are reduced due to the mass extraction in front of the turbine. With further increasing EGR rate, the losses due to flame extinction increase. The engine efficiency decreases again. For E85/i-C8, a continous decrease of engine efficiency can be noticed. Due to the increased specific heat, the probability of flame extinction increases at already low EGR rates.

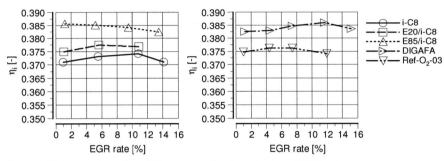

Figure 3-60: Indicated engine efficiency as a function of external EGR rate. $n_{mot} = 1500\ min^{-1}$, imep = 4 bar, stratified injection

This is verified when looking at Figure 3-61 displaying the standard deviation of indicated mean pressure as an indicator for cyclic fluctuations. Here, DIGAFA shows quite low cyclic fluctuations up to an EGR rate of twelve percent with $\sigma_{imep} < 0.1$ bar. Beyond that, the standard deviation increases rapidly.

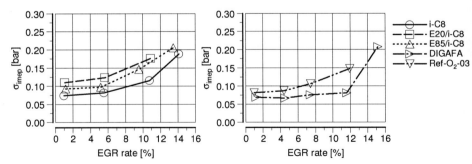

Figure 3-61: Standard deviation of indicated mean pressure as a function of external EGR rate. $n_{mot} = 1500\ min^{-1}$, imep = 4 bar, stratified injection

Looking at the dual-component fuels, i-C8 shows the lowest cyclic fluctuations with a continous increase with increasing EGR rate. E20/i-C8 shows the highest cyclic fluctuations, while, with further increasing ethanol fraction, slightly decreased cyclic fluctuations with E85/i-C8 can be observed. The effect of cyclic fluctuations on combustion efficiency can be explained by taking a closer look at the specific HC emissions in Figure 3-62. The multi-component fuels show at corresponding EGR rates lower HC emissions. DIGAFA shows the lowest HC emissions. In comparison to the dual-component fuels, the reference and the synthetic fuels consist of several hydrocarbon compounds improving the inflammability of the fuel [31]. Hence, the probability of flame extinction is reduced.

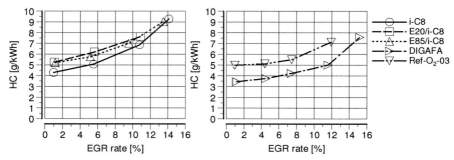

Figure 3-62: Specific HC emissions as a function of external EGR rate. $n_{mot} = 1500 \, min^{-1}$, imep = 4 bar, stratified injection

The specific CO emissions displayed in Figure 3-63 indicate the influence of EGR as well as ethanol addition on combustion temperature. With increasing EGR rate, the CO emissions increase due to reduced CO oxidation which is composed during dissociation. With high ethanol fraction (E85/i-C8), the highest CO emissions can be noticed while the multi-component fuels show the lowest CO emissions.

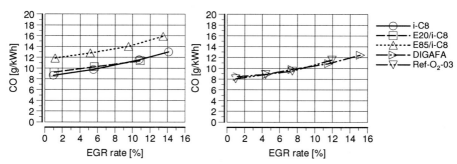

Figure 3-63: Specific CO emissions as a function of external EGR rate. $n_{mot} = 1500 \, min^{-1}$, imep = 4 bar, stratified injection

In antagonism to that, the NO_x emissions decrease with increasing EGR rate, shown in Figure 3-64. Looking at the fuel composition influence, a contrary behaviour can be found. With E85/i-C8, distinctively reduced NO_x emissions can be noticed. The NO_x emissions during engine run with E20/i-C8 are slightly higher in comparison to i-C8 as well as the NO_x emissions when using the multi-component fuels.

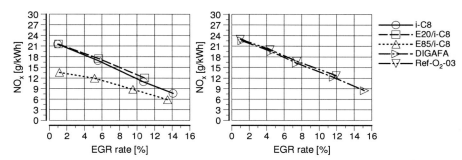

Figure 3-64: Specific NO$_x$ emissions as a function of external EGR rate. n_{mot} = 1500 min^{-1}, imep = 4 bar, stratified injection

This can be reasoned taking a closer look at the time of fifty percent mass fuel burned in Figure 3-65. Using ethanol enriched dual-component fuels, heat release takes place up to 1.5 to 2.5 °CA earlier resulting in increased NO$_x$ emissions for E20/i-C8.

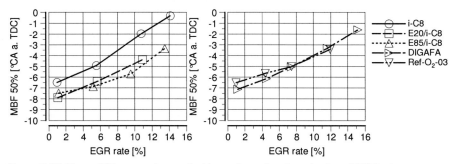

Figure 3-65: Time of fifty percent mass fuel burned as a function of external EGR rate. n_{mot} = 1500 min^{-1}, imep = 4 bar, stratified injection

Heat release deceleration due to increasing EGR rate is slighty lower when using the multi-component fuels. At maximum EGR rate, E85/i-C8 shows the lowest NO$_x$ emissions with lowest energy consumption. Considering multi-component fuels, the synthetic fuel shows the highest achievable EGR rate with high engine efficiency and NO$_x$ emissions as well as the highest combustion stability. All fuels do not emit detectable smoke emissions under this engine operation conditions.

Like shown in chapter 3.1.5, shifting injected fuel masses between the main injection impulses result in changes of heat release and combustion. While at high engine load with a low exhaust gas air-/fuel ratio a high injection split factor leads to increased engine efficiency, at low engine load, i.e. high exhaust gas air-/fuel ratio, this is limited by the enleanment of the mixture at the spark plug. The range of variation of the injection split factor describes the inflammability limits of fuels due to their link to the local spark plug air-/fuel ratio. Figure 3-66 shows the indicated engine efficiency for variations of the injection split factors using mono-/dual-component and multi-component fuels at standard engine operation point. All fuels show quite similar efficiency behaviour. With increasing injection split factor from the rich ignition limit on, engine efficiency develops quite constant up to μ_1 = 0.3. Beyond that, engine efficiency decreases until lean ignition limit is reached. Referring to the mono-/dual-component fuels, increasing engine efficiency can be found with increasing ethanol content like observed before. Ref-O_2-03 shows an engine efficiency on the level of i-C8, while DIGAFA reaches a level similar to E85/i-C8. Regarding the developing of the engine efficiencies, the ethanol enriched fuels show a distinctive maximum at μ_1 = 0.2, while i-C8 and the multi-component fuels develop quite constant at low injection split factors.

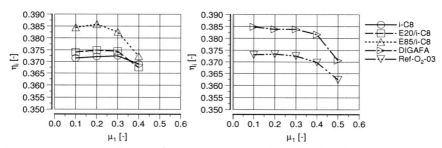

Figure 3-66: Specific indicated energy consumption as a function of injection split factor. n_{mot} = 1500 min^{-1}, imep = 4 bar, stratified injection

Furthermore, the ignition limit of the multi-component fuel can be found at a higher injection split factor of μ_1 = 0.5. This becomes evident when looking at the standard deviation of indicated mean pressure as an impression of the cyclic variations (Figure 3-67). I-C8 shows a low level of cyclic variations, while the ethanol enriched dual component fuel show a distinctive minimum. Before and beyond this minimum, cyclic variations increase. The multi-component fuels show a low level of cyclic variations over a varation range of 0.2. Here, DIGAFA leads to the lowest cyclic variations. Towards higher injection split factors, cyclic variations increase indicating the lean ignition limit.

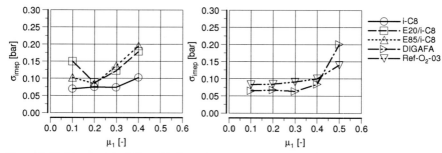

Figure 3-67: Standard deviation of indicated mean pressure as a a function of injection split factor. n_{mot} = 1500 min^{-1}, imep = 4 bar, stratified injection

The cyclic varations and therewith the stability of inflammation result in corresponding HC emissions in Figure 3-68. HC emissions during engine run with i-C8 are lower in comparison to the ethanol enriched fuels, especially towards higher injection split factors. Refering to the multi-component fuels, DIGAFA shows a considerable low level of HC emissions attesting a high inflammability limit during stratified engine operation.

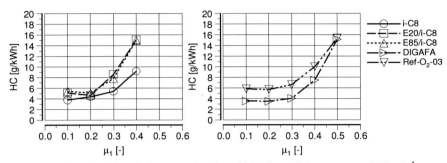

Figure 3-68: Specific HC emissions as a function of injection split factor. $n_{mot} = 1500\ min^{-1}$, imep = 4 bar, stratified injection

Figure 3-69 shows the measured specific CO emissions as a function of the injection split factor for the tested fuels. At low injection split factors, CO emissions remain constant for all fuels. With further increasing injection split factor, CO emissions increase. Ethanol addition yields to increasing CO emissions caused by the decreasing combustion temperature. The gradient of CO emissions increases with increasing ethanol content. The increasing cyclic variations and decreasing combustion temperature reduces CO oxidation resulting in higher CO emissions. Referring to the multi-component fuels, no distinctive difference can be observed.

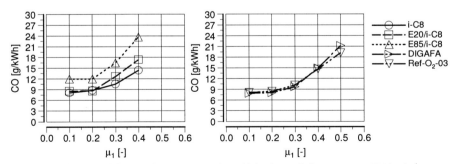

Figure 3-69: Specific CO emissions as a function of injection split factor. $n_{mot} = 1500\ min^{-1}$, imep = 4 bar, stratified injection

In agreement with discussed parameter variations, the NO_x emissions in Figure 3-70 show an antithetic behaviour compared to the CO emissions. With increasing ethanol content, NO_x emissions decrease. Also, with increasing injection split factor NO_x emissions decrease. The contradiction of increasing CO emissions and decreasing NO_x emissions points to a decreasing gas temperature with increasing ethanol content and towards higher injection split factors.

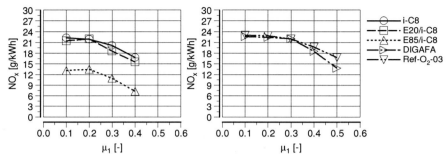

Figure 3-70: Specific NO_x emissions as a function of injection split factor. $n_{mot} = 1500\ min^{-1}$, imep = 4 bar, stratified injection

The injection split factor dependence can be reasoned by the developing of MFB 50% in Figure 3-71. Towards higher injection split factors, MFB50% is shifted to later positions resulting in decreased cylinder gas temperature. The developing of MFB50% reflects the developing of the CO emissions and the contradicted developing of the NO_x emissions.

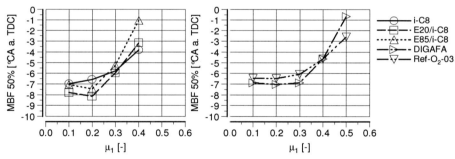

Figure 3-71: Time of fifty percent mass fuel burned as a function of injection split factor. $n_{mot} = 1500\ min^{-1}$, imep = 4 bar, stratified injection

Considering the inflammability limits of the mono-/dual-component fuels, it can be observed that with increasing ethanol fraction the variation range of injection split factor is getting smaller. Since multi-component fuels consist of several hydrocarbon boundings with different boiling behaviour, the ignition tolerance is quite higher in comparison to mono-/dual component fuels. Due to its higher evaporation enthalpy, ethanol addition diminishes ignition tolerance furthermore. Referring to Figure 3-67, all fuels show a similar degree of cyclic variations at $\mu_1 = 0.2$. When increasing or decreasing the injection split factor, cyclic variations increase when using the ethanol enriched fuels while i-C8 shows a quite stable developing. Both multi-component fuels show a quite stable inflammation over a wide range of injection split factor varation. In agreement to that, the mono-/dual-component fuels show a higher gradient of cyclic variations with increasing EGR rate. When reaching the lean ignition or EGR tolerance limit, cyclic varation increase strongly. DIGAFA shows the highest inflammation stability in regards to EGR tolerance as well as to mixture varation.

3.1.10   Efficiency analysis of ethanol enriched dual- and multi-component fuels

Figure 3-72 shows the calculated burn functions for the tested mono-/dual- and multi-component fuels as well as the times of fifty, seventy-five and ninety percent mass fuel burned (MFB 50%, MFB 75%, MFB 90%). Though only slight differences can be identified, MFB 90% is reached earliest using E85/i-C8. Refferring to the mono-/dual-component fuels, times of last twenty-five percent mass fuel burned are advanced with increasing ethanol fraction. Looking at the multi-component fuels, DIGAFA is converted faster during the last thirty percent of energy conversion.

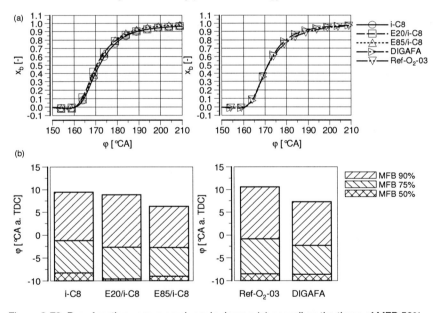

Figure 3-72: Burn function versus crank angle degree (a) as well as the times of MFB 50%, MFB 75& and MFB 90% (b) for dual-component (left) and multi-component (right) fuels. n_{mot} = 1500 min^{-1}, imep = 4 bar, λ = 2.8, μ1 = 0.2, IGA = 26.625 °CA b. TDC, stratified injection

Like observed in chapter 3.1.7, combustion temperature is distinctively influenced by fuel bounded oxygen enrichment. Figure 3-73 shows the burned gas temperature for the tested mono-/dual- and multi-component fuels at standard engine operation point. With increasing ethanol fraction, combustion temperature decreases due to increasing evaporation enthalpy and specific heat.

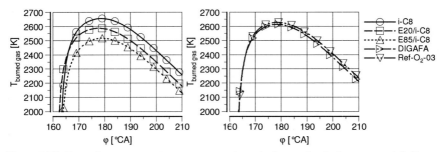

Figure 3-73: Burned gas temperature versus crank angle degree for dual-component (left) and multi-component (right) fuels. n_{mot} = 1500 min^{-1}, imep = 4 bar, λ = 2.8, μ_1 = 0.2, IGA = 26.625 °CA b. TDC, stratified injection

Looking at the multi-component fuels in the right diagram, combustion temperature is slightly decreased when using the oxygen enriched synthetic fuel (DIGAFA). The synthetic fuel is characterized by a slightly increased specific heat compared to the reference fuel.

Finally, splits of losses according to Weberbauer et al. [15] illustrate the influence of fuel modifications on engine efficiency (Figure 3-74). Based on the theoretical thermodynamic efficiency degree, losses due to effective compression ratio ($\Delta \eta_\varepsilon$) and under consideration of a reduced isentropic coefficient ($\Delta \eta_\kappa$) do not differ. Calculating a constant volume process with isenchoric heat release at the time of fifty percent mass fuel burned ($\Delta \eta_{MFB50\%}$), slightly increased losses using E20/i-C8 can be observed, while i-C8 shows the slightest loss. Determining the losses due to incomplete carburetion and combustion ($\Delta \eta_{HC+CO}$), the synthetic fuel DIGAFA shows considerable low losses. Losses when using the dual-component fuels show no differences. Considering the real engine process ($\Delta \eta_{RoHR}$), losses due to the rate of heat release decrease with increasing ethanol fraction when looking at the dual-component fuels. Referring to the multi-component fuels, no distinctive difference can be observed. Considering the real gas properties including gas dissociation, the use of ethanol addition in dual-component fuels results in increased engine efficiency. The caloric losses ($\Delta \eta_{Calorics}$) when operating with E20/i-C8 are slightest. Because of the decreased combustion temperature, E85/i-C8 results in lowest wall heat transfer losses ($\Delta \eta_{Wall\ heat}$), which finally reasons the distinctively increased engine efficiency. Referring to the multi-component fuels, caloric and wall heat losses do not differ decisively but are slightly higher than when using ethanol enriched dual-component fuels. The increased efficiency of the synthetic fuel is nearly exclusively caused by its more complete carburetion and combustion like it has been indicated by decreased cyclic fluctuations and higher inflammability limits.

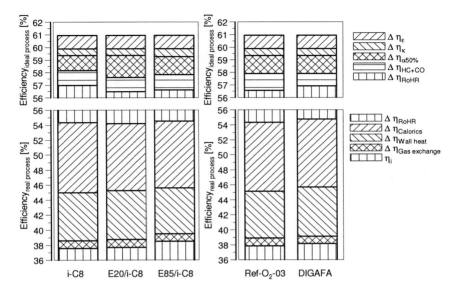

Figure 3-74: Split of losses for dual-component (left) and multi-component (right) fuels. $n_{mot} = 1500\ min^{-1}$, imep = 4 bar, $\lambda = 2.8$, $\mu_1 = 0.2$, IGA = 26.625 °CA b. TDC, stratified injection

Summary

Ethanol addition results in increased injection times due to the lower latent heat value of ethanol. This does not result in increased fuel penetration. Hence, no geometrical modifications for stratified carburetion using ethanol fuels are necessary. Using mono-/dual-component fuels, ethanol influence on non-premixed flame temperature is described. Non-premixed flame temperature decreases when adding ethanol because of an increased evaporation enthalpy and specific as well as absolute heat. Soot luminosity decreases distinctively with increasing ethanol fraction. At eighty-five percent ethanol fraction, no soot luminosity can be detected. This agrees with zero smoke emissions even at enriched mixture conditions. At standard engine operation condition, E85/i-C8 results in decreased NO_x emissions of about fourty percent. EGR tolerance is slighty lowered when adding ethanol, while homogenization tolerance is distinctively lower. Referring to engine efficiency, ethanol addition results in improved thermal efficiency. This is reasoned by decreased combustion temperature, i.e. decreased caloric and wall heat transfer losses, and faster heat release.

Multi-component fuels show an increased EGR and homogenization tolerance. Using synthetic fuel, smoke emissions are reduced. Due to more complete combustion, thermal efficiency of DIGAFA is similar to E85/i-C8.

3.2 The influence of EGR composition on engine knock

With increasing engine load the maximum cylinder pressure rises during the high pressure cycle. Depending on the fuel/air/residual gas mixture, consisting of burned and unburned components, the maximum pressure implicates a temperature threshold T_{thres}. Beyond this threshold, uncontrolled auto ignitions occur in the unburned endgas mixture, which can produce secondary pressure waves, which interfere with each other and reach engine damaging amplitudes.

Auto ignition phenomena are classified by Zeldovich [185] into

- deflagration,
- thermal explosion and
- developing detonation

depending on the temperature gradient in the unburned mixture. A deflagration is defined as auto ignition with a high gradient of temperature and low endgas temperature. A thermal explosion, which is comparable to a homogenous auto ignition, is described as auto ignition with low gradient of temperature and high endgas temperature. The remaining unburned gas is consumed abruptly with high flame speed. During a developing detonation, secondary pressure waves induced by several hot spots with a moderate gradient of temperature interfere with each other and generate a strong shock wave. Behind this shock wave, heat is released rapidly.

Considering the combustion chemistry described in chapter 2.2, these should be extended. The reaction paths of long chained hydrocarbons can be separated in three temperature ranges [32]. At high temperatures (T > 1200 K) the chain branching reaction of hydrogen radicals, reacting with oxygen to hydroxyl and oxygen radicals, are dominating:

$$\dot{H} + O_2 \rightarrow \dot{O} + \dot{O}H \tag{3-1}$$

Below this temperature, in the intermediate temperature regime, the hydroperoxid radical (HO_2) is mainly generated (3-2). By the abstraction of a hydrogen atom from an alkan, hydrogen peroxide (H_2O_2) is built (3-3), which, due to an interaction with a partner, decays into two hydroxyl radicals (3-4) and induces a chain branching reaction.

$$\dot{H} + O_2 + M \rightarrow \dot{H}O_2 + M \tag{3-2}$$

$$RH + H\dot{O}_2 \rightarrow \dot{R} + H_2O_2 \tag{3-3}$$

$$H_2O_2 + M \rightarrow 2\dot{O}H + M \tag{3-4}$$

Concerning endgas auto ignition in combustion engines the low temperature regime has to be considered mainly. The hydrogen peroxide decay (3-5) is slowed down and the alkyl peroxide radical (3-6) becomes dominant:

$$RH + O_2 \rightarrow \dot{R} + \dot{H}O_2 \text{ and} \tag{3-5}$$

$\dot{R} + O_2 \leftrightarrow R\dot{O}_2$. (first oxygen addition)   (3-6)

This alkyl peroxide radical can abstract hydrogen atoms under generating of hydroperoxid conjunctions (3-7). During the following chain branching reaction the hydroperoxid radical decays into an oxidation radical and a hydroxyl radical (3-8)

$$RH + R\dot{O}_2 \rightarrow ROOH + \dot{R}$$ (3-7)

$$ROOH \rightarrow R\dot{O} + \dot{OH}$$ (3-8)

Further on, after an internal hydrogen abstraction, a second oxygen addition is induced following the mechanism reported by Chevalier [186] and leads to a two-stage ignition, which is characteristic for low temperature regime auto ignition. The precursors of the chain branching reaction during the oxygen addition decay. A degenerated chain branching reaction occurs, leading to a brake down of the chain branching with a low temperature increase. After an ignition delay time, a second ignition occurs with complete reaction in the high temperature regime. Accordingly, due to the degenerated chain branching reaction, the ignition delay is increased in spite of a temperature increase, leading to a negative temperature coefficient (NTC). The temperature range of NTC characterizes the migration of the low to the intermediate temperature regime with the competing occurrence and consumption of alkyl and hydro peroxides. Furthermore, the NTC progress of hydrocarbon/oxygen mixtures characterizes their auto ignition behaviour. Griffiths et al. [201] have investigated the impact of the NTC by variation of the compression end temperature using a rapid compression machine. If compression temperatures are reached in the region of positive temperature coefficients, the reactions are initiated in the hottest endgas zones. In opposition to that, if compression temperature is in the region of negative temperature coefficients, an increased reactivity can be observed in the colder endgas zones. This increased production of intermediate products and radicals leads to knocking combustion originated in the coldest endgas zone.

Pöschl scetched the qualitative pressure and temperature developing during low temperature auto ignition and classified the ignition delay time IDT in three periods (Figure 3-75, [187]):

- IDT_1: Peroxide formation;
- IDT_2: Cold flame occurrence caused by peroxide decay and aldehyde formation;
- IDT_3: Blue flame occurrence caused by aldehyde decay and first CO formation.

Finally, the ignition delay time IDT is defined as time span between first reaction start and the occurrence of CO_2 as combustion product. It will be used for determining influences on auto ignition tendency of cylinder mixtures using zero-dimensional reactor simulations (compare appendix h).

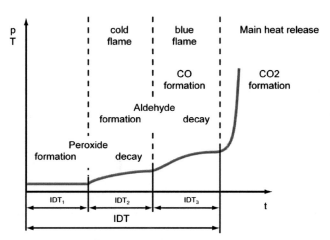

Figure 3-75: Qualitative pressure and temperature curve during the multi-stage low temperature ignition. Taken from Pöschl [187].

3.2.1 Character of engine knock

Uncontrolled spontaneous auto ignition of unburned gas, or endgas, in internal combustion engines is mainly described as spark knocking. According to Heywood [31], the spark knocking needs to be separated from abnormal combustion phenomena resulting from surface ignition. Here, the unburned gas mixture is ignited by overheated parts of the combustion chamber. This surface ignition cannot be controlled by engine control. Heywood [31] developed a classification of abnormal combustion phenomena which has been extended by Schintzel et al. [188].

Several investigations have been carried out considering the occurrence of spark knocking. Following the auto ignition theory, spark knocking arises due to a too high compression of the endgas and can be controlled via the ignition time. This compression is caused by the propagating primary flame front. With increasing compression the endgas temperature increases and the fuel oxidation process starts and leads to local auto ignition (thermal explosion). Due to this auto ignition, further endgas zones are compressed and ignite the same way extremely rapidly [31]. During spark knocking pressure waves with a frequency higher than five kHz occur [189].

In opposition to the auto ignition theory, Pöschl [187] and Heywood [31] give references about authors reporting about a detonation theory. Spark knocking combustion showed rapid flame propagation and no thermal explosions. Several authors approve that, when considering temperature and mixture inhomogenities or exothermic centres, detonation can arise in combustion engines [190]-[194].

Hence, engine knock can be characterized as combinations of local thermal explosions and detonations depending on temperature and mixture inhomogenities induced by exothermic centres [34],[195]. From these on, the secondary flame front propagates deflagratively. Depending on the endgas condition this secondary flame front can be accelerated and migrate into a developing detonation [18], which results in a considerable higher knock intensity. The migration to a developing detonation is described in detail by Pöschl [187].

Due to nearly adiabatic conditions for the endgas during combustion, the endgas temperature can change between the low temperature and the intermediate temperature range. The formation of the alkyl peroxide and the hydro peroxide radical compete against each other. During this state, high hydroxyl radical concentrations have been observed in several investigations [196]-[198] and can be linked to the occurrence of a cold and a blue flame initiating a two-stage ignition, referring to chapter 2.2. Authors of further publications use the occurrence of formaldehyde (CH_2O) during cool flame appearance for the detection of hot spots [199],[200]. The transition from deflagrative combustion to knocking combustion occurs at temperatures of minimum ignition delay time [187].

Summing up, endgas auto ignition, i.e. spark knocking, is induced by exothermic centres [195], which lead to low temperature regime reaction path with a cold and blue flame. At this point already pre-reactions have taken place whose intensity is influenced by the degree of oxidation of residual components. After ignition, the exothermic centres increase the endgas temperature even more and enrich further the endgas with reactive intermediate products. The following propagation depends on the local distribution of exothermic centres and the temperature gradient in the endgas. High endgas temperature gradients lead to an increased rate of heat release, which can migrate to a developing detonation with high amplitude secondary pressure waves.

3.2.2 Knock suppression in turbo charged gasoline engines

Next to the fundamentals of spark knocking, other influences on engine knock has to be considered for greatest possible knock suppression during engine run. Like described before, the occurrence of spark knock is directly linked to the in-cylinder pressure, the resultant gas temperature and the unburned gas mixture. Hence, spark knocking limits the maximum compression ratio of a spark ignition engine and therewith its maximum thermodynamic efficiency.

In today's turbo charged gasoline engines, the ignition time is delayed to assure a certain avoidance of uncontrolled auto ignition during engine run. This leads to a increase of exhaust gas temperature, which endangers the durability of engine parts in the exhaust path. Therefore, additional fuel is induced into the combustion chamber reducing the exhaust gas temperature due to its higher specific heat capacity. Fuel consumption increases much. To avoid this fuel enrichment and improve combustion efficiency, auto ignition tendency in the unburned mixture has to be lowered. Thus, optimal spark knock suppression is desirable and can be improved by

- fuel modification,
- combustion velocity enhancement and
- cylinder charge modification.

The use of more auto ignition resistant fuels leads to reduced auto ignition tendency. This auto ignition resistance of fuel is mainly described with the research octane number (RON) and the motor octane number (MON). Detailed descriptions of auto ignition tendency of hydrocarbons are given by Spausta [203] or Lovell [202].

Referring to page 19f., the increasing of cylnder charge turbulence leads to higher turbulent flame velocity. Unburned mixture is consumed before auto ignition occures. The increased turbulence and therewith accelerated flame propagation increases the danger of developing detonations, if auto ignition occurs in the endgas. Ganser [204] for example reports that with increasing turbulence the primary flame propagation is accelerated and the auto ignition tendency under constant boundary conditions is lowered. When increasing engine load or shorting the spark ignition delay, spontaneous endgas auto ignitions with high pressure amplitudes occur. Besides increasing charge turbulence, combustion velocity enhancement can be achieved via gas modification. With adding of Hydrogen (H_2) and carbon monoxide (CO) the combustion process is accelerated due to enhanced reactivity. The flame reaches the unburned endgas mixture soon enough, so there is not enough time for auto ignition occurrence [206]-[211].

The occurrence of spark knock can be suppressed by modification of unburned mixture's caloric condition. To lower the effect of charge heating due to residual gas on the one hand and to use the effect of increased specific heat capacity on the other hand, external recirculation of cooled exhaust gas (EGR) is a promising approach to increase ignition delay time of the unburned mixture [207],[212]-[217]. The increased specific heat capacity leads to decreased compression and combustion temperature [31]. The ignition delay time of the unburned mixture is increased. The spark ignition time delay can be reduced, the thermodynamic efficiency of the combustion process is improved. Important to notice, that the advantage of decreased auto ignition tendency outweighs the disadvantage of lowered laminar flame speed due to the increased specific heat capacity. With an increase of turbulence intensity this can be partially countervailed [213].

Next to the known investigations concerning the thermodynamic effect of external cooled EGR during high load combustion, the thermo kinetic effects of EGR are mostly neglected. Referring to chapter 3.2.1 and its reflections on the influence of pre-conditioning the gas mixture before auto ignition occur, Green et al. [205] have reported that completely oxidized components decrease and partially oxidized components increase pre-reaction's intensity. Thus, the effect of EGR composition on auto ignition should be considered. Because of incomplete combustion and high combustion temperature, the EGR composition (see appendix Figure F-7-8) in a DISI-Engine shows significant amounts of unburned HC and NO.

The influence of unburned or partially burned HC on auto ignition is in detail discussed by Heywood [31], Lovell [202] and Spausta [203]. Former publications also already discuss the influence of increased NO concentrations in fresh gas mixtures on knock occurrence. Prabhu et al. [218] and Kawabata et al. [219] found a direct link between the NO concentrations in the unburned gas mixture and knock occurrence during engine run. Stenlåås et al. [220] numerically investigated the reaction paths leading to increased knock occurrence driven by NO. Depending on the temperature range, the main initiating reaction during intermediate temperature oxidation leading to earlier auto ignition is

$$NO + H\dot{O}_2 \rightarrow NO_2 + \dot{O}H \qquad (3\text{-}9)$$

During low temperature oxidation composed alkylperox radical (RO_2) (compare reaction 3-6) reacts with nitric oxide to nitric dioxide (NO$_2$) under formation of alkyperox radicals (RO)

$$NO + R\dot{O}_2 \rightarrow NO_2 + R\dot{O}$$ (3-10)

At higher NO concentrations, a third nitric oxide driven reaction occurs

$$NO + \dot{O}H + M \rightarrow H\dot{O}NO + M$$ (3-11)

Knock occurrence is therefore directly linked to the NO concentration and the compression temperature. A lower start temperature and high NO concentrations decelerate the reaction. The formed OH radicals are consumed for the formation of azotic acid (HONO), following reaction path (3-11). At lower NO concentration, OH radicals are composed via reaction path (3-10) much faster than they can be consumed on path (3-11). Thus, the reactivity of the unburned gas is increased. With higher start temperature, the NO concentration threshold is moved to higher NO concentrations. Additionally, with increasing start temperature the influence of the intermediate temperature oxidation path (3-9) increases due to a distinctive NTC behavior. Reaction path (3-11) needs much more NO for consuming via path (3-9) and (3-10) composed radicals.

Summarizing, the increased NO concentration in the unburned mixture, due to recirculation of partially oxidized exhaust gas, decelerates the chemical conversion, because NO consumes hydroxyl radicals composing HONO. Due to that, the endgas reaches earlier the decay limit of hydroxyl peroxid to hydroxyl radicals and ignites. Referring to Prabhu et al. [218], this effect is quite distinctive at the use of external EGR during stoichiometrical engine run.

3.2.3 Study on engine knock tendency

To investigate the effect of NO$_x$ and unburned HC on auto ignition in an internal combustion engine, EGR will be converted using the three-way-catalyst mechanism:

$$2CO + O_2 \rightarrow 2CO_2$$ (3-12)

$$C_xH_y + \left(x + \frac{y}{4}\right)O_2 \rightarrow xCO_2 + \left(\frac{y}{2}\right)H_2O$$ (3-13)

$$2NO + 2CO \rightarrow 2CO_2 + N_2 \text{ and}$$ (3-14)

$$2NO + 2H_2 \rightarrow 2H_2O + N_2$$ (3-15)

Thereby, the engine exhaust gases are reformatted to ideal EGR according to the previously described assumptions. Due to the nearly completely elimination of NO$_x$ and unburned HC in the cylinder mixture a reduced tendency to endgas auto ignition is expected when running the engine with the EGR catalyst. The influence of the fuel-/air ratio and its variation caused by different fuel enrichment need has to be considered as well. A shortcut of the investigations is given in [221].

Appendix Figure B-7-4 shows the engine setup as well as the position of the temperature and pressure measurement points on the intake and exhaust side. The EGR route of the engine is complemented with a standard upstream three-way-catalyst based on a Palladium/Rhodium-Ratio of 9:1 right after the exhaust manifold.

The EGR route is built up on modular parts, allowing a quick exchange of the EGR catalyst with the standard EGR pipe. Exhaust emissions are sampled just behind the turbine exit and analyzed using an SESAM-3 system [222] and an AVL SmokeMeter [223]. Temperature and pressure measurement points are set in front of and behind the catalytic converter and the EGR cooler (appendix Figure B-7-4). The EGR is sampled in front and behind of the catalyst where the EGR component concentrations are measured. In this way, the EGR catalyst conversion rate can be controlled. Moreover, these measured concentrations are used as initial concentrations for numerical simulations. Numerical studies are performed using a zero-/one-dimensional reactor network (comp. appendix k) imitating the engine setup. Appendix Figure H-7-9 shows the used reactor network consisting of a homogenous reactor imitating the combustion chamber with constant volume. The EGR cooler and the adjacent EGR pipe are modeled using plug flow reactors (PFR) and the throttle as mixer composing the combustion mixture. The reactor network is fed by pressure and temperature data of the corresponding test bench investigation. EGR composition and concentration is taken from test bench investigations. With this, the reactor network is initialized. EGR is cooled down in the plug flow reactors considering low temperature reactions like the conversion of NO to NO_2, for example. Then, EGR is mixed with fuel and fresh air given by test bench investigation data and fed into the homogenous reactor imitating the combustion chamber. The homogenous reactor is initiated with temperature and pressure at ignition time taken from detailed thermodynamic analysis. The system evolves until equilibrium conditions are reached. The ignition delay time is defined by the occurrence of the maximum gradient of CO_2 concentration.

For investigating the effect of the recirculated modified exhaust gas on high-load combustion, the amount of EGR is varied at constant load with constant fuel-/air ratio (ϕ) as well as constant exhaust gas temperature at 80 % of maximum brake torque (MBT) and two different engine speeds (Table B-7-3). The ignition time is chosen using a standard knock criterion [224] calculated from the cylinder pressure (see appendix Figure C-7-6). Additionally, variations of the air-/fuel ratio are performed to differentiate between mixture leaning using EGR or oxygen.

The influence of EGR on engine knock

Figure 3-76 shows the reached ignition angle for an EGR variation at constant engine speed and load. Like expected the ignition angle can be continously increased, i.e. shifted to earlier position, with increasing external EGR rate. Likewise the calculated time of fifty percent mass fuel burned as a function of EGR rate is displayed. It varies with ± 1 crank angle degree, but does not differ definitely with increasing external EGR rate. With increasing EGR rate, laminar flame velocity is lowered due to the increased specific heat and therewith lowered gas temperature.

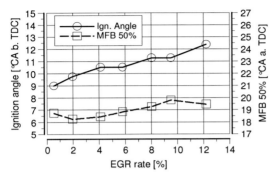

Figure 3-76: Reached ignition angle and time of fifty percent mass fuel burned (MFB 50%) for an EGR variation. $n = 2500\ min^{-1}$, imep = 17.5 bar, $\lambda = 1$

Consequently, a decreased rate of heat release can be observed in Figure 3-77. Because of earlier ignition an ealier rise of heat release can be observed with EGR = 12%.

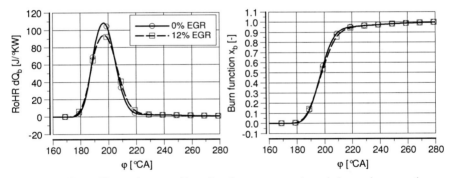

Figure 3-77: Rate of heat release and burn function versus crank angle for engine operation with closed EGR valve and an EGR rate of twelve percent. $n = 2500\ min^{-1}$, imep = 17.5 bar, $\lambda = 1$

The rate of heat release is lower and therewith results in a later heat release in the end. Like it can be seen in the right diagram looking at the burn function for both operating points, the time of fifty percent mass fuel burned (MFB 50%) is quite equal.

The measured indicated engine efficiency η_i in Figure 3-78 displays a slight increase with increasing EGR rate while the standard deviation of indicated mean pressure σ_{imep} increases. The increased fraction of inert gas components destabilizes ignition and increases the probability of flame extinction [31].

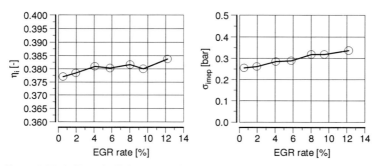

Figure 3-78: Indicated efficiency η_i and standard deviation of indicated mean pressure σ_{imep} as a function of EGR. $n = 2500\ min^{-1}$, imep = 17.5 bar, $\lambda = 1$

Consequently, increased HC emissions can be observed in Figure 3-79 in the right diagram. On the other hand, CO emissions slightly decrease with increasing EGR rate despite a decreasing combustion temperature. In the left diagram, nitric oxide NO and soot emissions are shown as a function of EGR. NO_x emissions decrease linearly with increasing EGR rate while soot emissions increase. While NO_x composing is directly linked to combustion temperature, the increasing soot emissions cannot be explained definitely. Next to heterogeneous mixture due to unsatisfying carburetion, fuel spray-wall interaction plays an important role. Here, the interaction between the cone shaped fuel spray and the opened inlet valves is found to be the main soot source [12]. It is assumed that the lowered combustion temperature reduces soot oxidation and consequently results in increased soot emissions with increasing EGR rate.

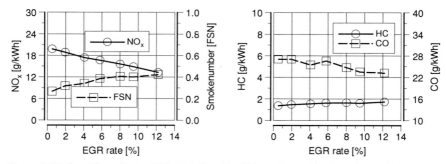

Figure 3-79: Specific emission (SE) of nitric oxides NO_x, unburned hydrocarbons HC, carbon monoxide CO and smokenumber as a function of EGR. $n = 2500\ min^{-1}$, imep = 17.5 bar, $\lambda = 1$

In summary, a split of losses (Figure 3-80) illustrates the influence of EGR on engine operation. Due to the increased residual gas fraction in the cylinder, efficiency of the ideal process considering the mixture composition is lowered ($\Delta\eta_k$) when running the engine with EGR.

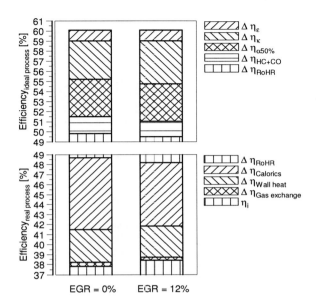

Figure 3-80: Split of losses for engine operation without EGR and with twelve percent EGR. $n = 2500\ min^{-1}$, imep = 17.5 bar, $\lambda = 1$

Considering the real process and its heat release developing, $\Delta\eta_{RoHR}$ is little higher comparing to engine operation without external EGR. External EGR reduces the mass flow through the turbo chargers turbine and reduces exhaust back pressure. Gas exchange losses are reduced. Finally, the main differences can be found in $\Delta\eta_{Calorics}$ and $\Delta\eta_{Wall\ heat}$. The decreased combustion temperature due to increased specific heat at EGR = 12% results in considerable efficiency increase. Figure 3-81 shows the corresponding curves of gas mean temperature for engine operation without and with twelve percent EGR. During heat release, a considerable temperature offset of up to 200 K can be observed corresponding to the discussed lower losses and decreased NO_x emissions. Wall heat transfer is reduced as well as calorical losses due to gas dissociation.

Regarding auto ignition tendency the gas temperature during compression is essential. The right diagram shows a section of the temperature developing. At twelve percent EGR a quite lower compression temperature is found. Considering earlier ignition time a considerable decrease of gas temperature at ignition time T_{IGA} is found.

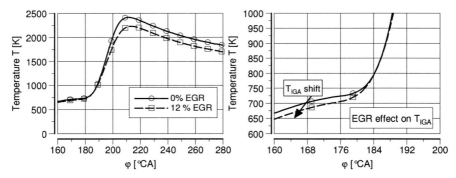

Figure 3-81: Gas mean temperature T during high-pressure cycle without EGR and twelve percent EGR displaying EGR effect on in-cylinder temperature. $n = 2500\ min^{-1}$, $imep = 17.5\ bar$, $\lambda=1$

Figure 3-82 shows the gas temperature T_{IGA} and the corresponding cylinder pressure p_{IGA} at ignition time. Due to constant engine load, p_{IGA} slightly decreases with shifting ignition time to earlier position. With increasing EGR rate a nearly linear decrease of T_{IGA} can be observed while specific heat at ignition time remains constant. Considering the diluting effect of EGR [31], the absolute heat of the in-cylinder gas increases resulting in decreasing temperature.

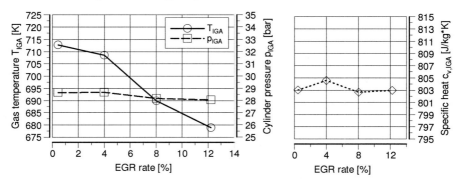

Figure 3-82: Gas temprature T_{IGA}, cylinder pressure p_{IGA}, and specific heat $c_{v,IGA}$ at ignition angle for EGR variation. $n = 2500\ min^{-1}$, $imep = 17.5\ bar$, $\lambda= 1$

Initializing the zero-/one-dimensional reactor network with p_{IGA} and T_{IGA} leads to a theoretical ignition delay time and EGR's influence on it. Figure 3-83 shows the already described ignition angle and the calculated ignition delay time (IDT) for this EGR variation. EGR composition is taken from EGR analysis using the test bench exhaust gas analyzer.

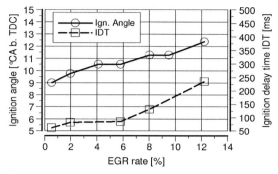

Figure 3-83: Achieved ignition angle and calculated ignition delay time (IDT) for an EGR variation. *n = 2500 min⁻¹, imep = 17.5 bar, λ= 1*

Up to six percent EGR, a tendency to increased IDT can be observed. Beyond six percent EGR, IDT increases linearly with increasing EGR rate. The good agreement of ignition angle and IDT verifies the link between engine test bench and a zero-/one-dimensional reactor network.

Thus, to gain general understandings of kinetical influence of EGR on engine knock, the numerical study is extended using multi-reactor systems. By this, fluctuations of air-/fuel ratio and temperature are considered. These fluctuations present the initial conditions for each single-reactor-network, which develop temporally self-dependent until ignition occures. Using a MATLAB-Tool developed by Mauss [225], the calculated ignition delay times are plotted as λ-T-maps, also published as kinetic mapping [226].

Figure 3-84 shows kinetic maps of ignition delay time (IDT) for a constant pressure of p = 20 bar with no EGR (left) and EGR = 30 % (right). EGR compostion is taken from engine measurements at an EGR rate of EGR = 8 %.

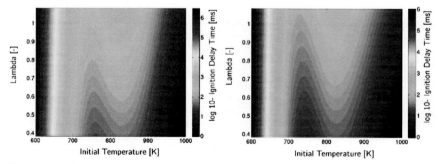

Figure 3-84: Kinetic maps with EGR = 0% (left) and EGR = 30% (right). *p = 20bar, RON 95 (95% iso-octane / 5% n-heptane)* [221]

Both maps show quite similar characteristics. Below 650 K, ignition delay time depends mainly of temperature. Beyond 650 K a strong influence of the air-/fuel ratio can be observed, displaying a distinctive NTC behaviour up to 850 K. Adding EGR, a decrease of calculated ignition delay time is found. The effect of EGR is most distinctive in the intermediate temperature regime resulting also into a shift of the NTC range to lower temperatures. Regarding temperature developing during engine operation (comp. Figure 3-82), the in-cylinder mixture is shifted to lower temperatures with increasing EGR rate. Looking at the kinetic maps, these result in increasing IDT. Due to temperature decrease, IDT increases despite the IDT decreasing effect of EGR. Especially towards higher EGR rates (in this case beyond EGR = 6 %, compare Figure 3-83), the temperature effect seems to get more and more important.

Next to stoichiometrical engine operation, the use of external cooled EGR is a quite promising approach to lower the need for fuel enrichment. Due to EGR, specific and absolute heat increases and gas temperature decreases. Additionally, earlier ignition can result in earlier combustion. This all result in a lowered exhaust gas temperature and a decreased need for fuel enrichment. Looking at Figure 3-84, the effect of decreased temperature regarding IDT and therewith knock tendency is lowered. Reducing fuel enrichment leads to leaner in-cylinder mixture. The engine operation point in the kinetic map shifts to the upper left instead horizontally to the left, resulting in lower IDT values. Additionally, the reduced fuel enrichment results in a decreased specific heat at constant EGR rate. The temperature decrease is derogated.

Figure 3-85 shows ignition angle and IDT for an EGR variation at constant engine speed and load. In opposition to the first discussed EGR variation, the exhaust gas temperature in front of the turbine T_3 is limited to $T_3 = 1173$ K.

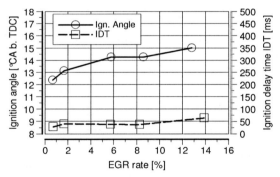

Figure 3-85: Achieved ignition angle and calculated ignition delay time (IDT) for an EGR variation. $n = 3000\ min^{-1}$, $imep = 19\ bar$, $T_3 = 1173\ K$

Because of this, the engine needs to be operated with a fuel enriched mixture varying between $0.91 \leq \lambda \leq 0.95$. With increasing EGR rate, the fuel enrichment can be reduced because of decreasing exhaust gas temperatures. A continuos shift to earlier ignition can be observed with increasing EGR rate. Regarding the developing IDT, a slight increase can be identified at higher EGR rates. Generally, the gradient of IDT is much lower comparing to Figure 3-83 and there is no clear correlation between ignition angle and IDT. The influence of fuel enrichment on IDT seems to be most important for this engine operating condition. Regarding initalizing T_{IGA} and the corresponding specific heat taken from test bench measurements, increasing EGR rate leads to a linear decrease of T_{IGA} of about 30 K comparable to the stoichiometrical case. Specific heat increase correlates directly with increasing λ.

Comparing to Figure 3-82 with stoichiometrical mixture, fuel enrichment dominantly influences caloric properties of the compression gas while temperature decrease can be linked to the reduced compression pressure at ignition time. Thus, the influence of ignition angle on initial conditions is unneglectable.

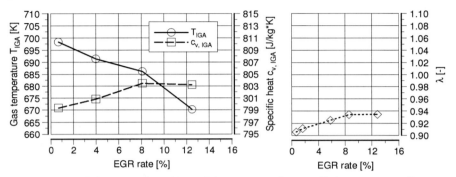

Figure 3-86: Initializing temperature T_{IGA} and the corresponding specific heat $c_{v,IGA}$ as well as air-/fuel ratio λ as a function of the EGR rate. $n = 3000 \ min^{-1}$, $imep = 19 \ bar$, $T_3 = 1173 \ K$

In summary, an increase of ignition delay time, and therewith possible earlier ignition, like it is found during engine operation, is nearly exclusively reasoned by the decrease of compression end temperature. Regarding kinetic maps, fuel enleanment countervails the EGR effect on auto ignition tendency. During enriched engine operation this cannot distinctively be observed on the test bench. Furthermore, the kinetic maps show an auto ignition enhancing effect under consideration of on-engine measured EGR composition.

Referring to theoretical investigations described in chapter 3.2.2, a reformation using the three-way-catalyst mechanism of EGR to ideal EGR only consisting of CO_2 and H_2O seems to be worthwile. Under assumption of quite similar temperature behaviour during compression, a distinctive decrease of knock tendency during engine run is expected regardless of given air-/fuel ratio.

The influence of EGR reformation using an EGR catalyst

In order to aim optimal conversion of the catalytic converter, a stoichiometric fuel-/air ratio is needed. Additionally, a specific exhaust gas temperature range is necessary during conversion. Here, the exhaust gas temperature is set between $T_3 = 1073 \ K$ to 1173 K, depending on the EGR rate and operating point. Hence, conversion rate of the EGR catalyst can be surveilled. Figure 3-87 shows the conversion rates C_{Cat} of the EGR catalyst as function of EGR rate during stoichiometrical engine operation.

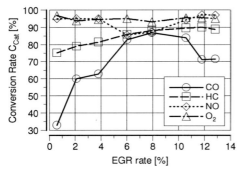

Figure 3-87: Conversion rates of EGR catalyst C_{Cat} for conversion of CO, HC, NO and O_2 as a function of EGR rate. *n = 2500 min^{-1}, imep = 17.5 bar, λ= 1*

For conversion of HC's quite high conversion rates are achieved while the conversion of CO is rather poor. This can be explained by the simple air-/fuel ratio control used during these investigations. Using a standard air-/fuel ratio control, a cyclic variation of the fuel flow at a frequency of 0.5 to 1 Hz is impressed to ensure a high conversion efficiency of the catalyst [31]. Here, a simple fuel flow control is used without any impressed cyclic variations using one broad band lambda sensor in front of the main catalyst.

With further opening the EGR valve and an increasing mass flow through the EGR catalyst, CO conversion increases, but is still quite low with conversion rates of sixty to seventy percent. HC's conversion increases continuously with increasing EGR mass flow, while NO and oxygen are nearly completely converted.

It needs to be pointed out, that the carbon dioxide (CO_2) concentration increases due to catalysis (compare reactions (3-12) – (3-14)). This is important to keep in mind, since the external EGR rate is calculated through the measured CO_2 concentration in the intake manifold. The catalytic process introduces an error in calculating the external EGR rate, which is calculated to be less then 2 %.

Figure 3-88 shows the achieved ignition angle at knock limit and the resulting time of fifty percent mass fuel burned for an EGR variation with and without EGR catalyst. In comparison to the discussed EGR variation, using the EGR catalyst a further increase of two to three degree crank angle regarding the achievable ignition angle can be observed. The earlier ignition time results in a corresponding earlier time of fifty percent mass fuel burned.

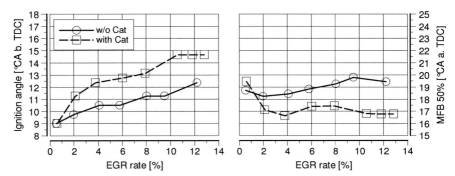

Figure 3-88: Achieved ignition angle at knock limit and the resulting time of fifty percent mass fuel burned (MFB 50%) as a function of EGR with and without EGR catalyst. $n = 2500\ min^{-1}$, $imep = 17.5\ bar$, $\lambda = 1$

As shown in Figure 3-89, time and shape of heat release is shifted to earlier time with similar developing. Consequently, EGR modification allows an earlier time of combustion but does not have a distinctive effect on rate of heat release.

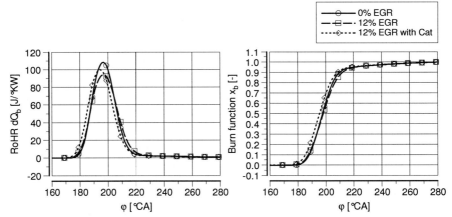

Figure 3-89: Rate of heat release and burn function without EGR, with twelve percent EGR and with twelve percent EGR plus EGR catalyst. $n = 2500\ min^{-1}$, $imep = 17.5\ bar$, $\lambda = 1$

Next to decreased knock tendency, EGR modification results in a continuously increased engine efficiency and decreased standard deviation of indicated mean pressure in comparison to engine operation with standard EGR (Figure 3-90). The shift of heat release to earlier time increases combustion pressure and temperature and consequently stabilizes combustion. Less flame extinction can be assumed and cyclic variations are reduced.

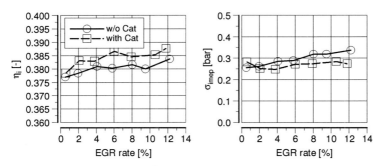

Figure 3-90: Indicated engine efficiency η_i and standard deviation of indicated mean pressure σ_{imep} as a function of EGR with and without EGR catalyst. $n = 2500 \, min^{-1}$, $imep = 17.5 \, bar$, $\lambda = 1$

In Figure 3-91, comparable HC emissions can be observed with increasing EGR rate when using the EGR catalyst. Regarding CO emissions, these are continously a little lower during engine operation with EGR catalyst. The earlier time of combustion let expect a higher combustion temperature and therewith a higher CO oxidation rate.

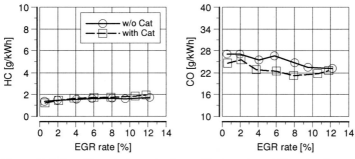

Figure 3-91: Specific emissions of unburned hydrocarbons HC and carbon monoxide CO as a function of EGR rate with and without EGR catalyst. $n = 2500 \, min^{-1}$, $imep = 17.5 \, bar$, $\lambda = 1$

NO_x emissions show for both engine operation modes a clear decrease with increasing EGR rate (Figure 3-92). In opposition to the assumption of increased combustion temperature like CO emissions let purpose, NO_x emissions with EGR catalyst are continously a little bit lower. Here, the catalytic mechanism of the EGR catalyst has to be kept in mind. When using the EGR catalyst, oxygen in the EGR is consumed during oxidation of CO and HC. Hence, the oxygen content in the cylinder is lowered reducing NO composing in comparison to the standard EGR operation mode despite increased combustion temperature. Additionally, NO concentration in the EGR is reduced and less NO is recirculated.

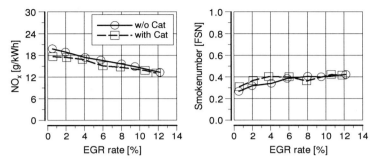

Figure 3-92: Specific emissions of nitric oxides NO$_x$ and filter smoke number FSN as a function of EGR rate with and without EGR catalyst. *n = 2500 min^{-1}, imep = 17.5 bar, λ= 1*

The smoke emissions in Figure 3-92 (right diagram) increase degressively with increasing EGR rate. In comparison no effect of the EGR catalyst can be observed.

Finally, a split of losses in Figure 3-93 presents the main reasons for increased engine efficiency.

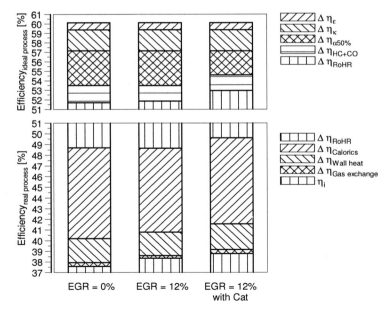

Figure 3-93: Split of losses when operating the engine without EGR, with twelve percent EGR and with twelve percent EGR plus EGR catalyst. *n = 2500 min^{-1}, imep = 17.5 bar, λ= 1*

Next to small differences regarding in-cylinder mixture composition ($\Delta\eta_\kappa$), the earlier time of heat release decreases dominantly thermodynamic losses when using the

EGR catalyst. Otherwise, when using the EGR catalyst slightly increased losses occur due to caloric effects, wall heat transfer and gas exchange. Gas temperature curves displayed in Figure 3-94 explains the increased losses.

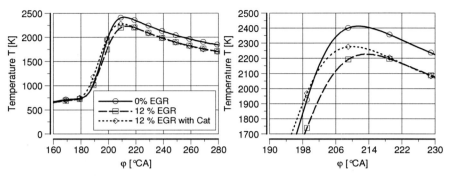

Figure 3-94: Gas mean temperature without EGR, with twelve percent EGR and with twelve percent EGR and EGR catalyst. $n = 2500 \ min^{-1}$, imep = 17.5 bar, $\lambda = 1$

The earlier time of heat release results in higher cylinder pressure and consequently higher temperatures like it has been purposed at emission discussion. Hence, more dissociation and wall heat transfer occur. Efficiency increase due to earlier heat release is slightly countervailed by increased losses due to the resulting higher gas temperature. Increased gas exchange losses when using the EGR catalyst are explained by increased pressure losses over the catalyst in the EGR route.

Based on detailed thermodynamic data, initial conditions for calculation of the theoretical ignition delay time are determined. Like discussed before, regarding external EGR variations IDT dominantly depends on the initial temperature T_{IGA}. Figure 3-95 shows the determined temperatures T_{IGA} with and without EGR catalyst. Despite the earlier ignition time when using the EGR catalyst (comp. Figure 3-88), T_{IGA} is quite similar to engine run without EGR catalyst confirming the prior assumption of equal initial condition (page 125).

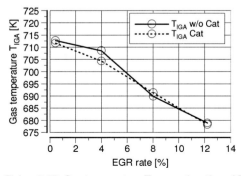

Figure 3-95: Gas temperature T_{IGA} as a function of EGR with and without EGR catalyst. $n = 2500 \ min^{-1}$, imep = 17.5 bar, $\lambda = 1$

Comparing the determined IDT with the achieved ignition angle at knock limit (Figure 3-96), a qualitative correlation can be observed. Like in Figure 3-83, up to six percent, a slight increase of IDT with increasing EGR rate can be seen. Beyond six percent EGR, IDT increases exponentially. Comparing initial condition taken from engine run with and without EGR catalyst, calculated IDT increases considerably stronger when using EGR catalyst. This correlates with reduced knock tendency during engine operation.

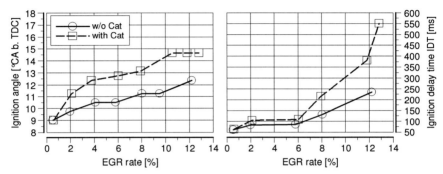

Figure 3-96: Achieved ignition angle at knock limit and calculated ignition delay time (IDT) as a function of EGR with and without EGR catalyst. $n = 2500\ min^{-1}$, imep = 17.5 bar, λ= 1

Since initial thermodynamic conditions are shown to be quite equal, increased IDT can be distinctively reasoned by the modified EGR composition and approve the postulated influence of EGR composition on auto-ignition tendency.

With regard to efficiency increase meaning decrease of fuel consumption, engine operation with EGR shows quite higher potential when reducing additionally the need for fuel enrichment at higher engine speeds. Due to the increased specific heat not only knock tendency but also exhaust gas temperature is lowered. Consequently, less fuel enrichment is needed for exhaust component protection. Therewith, the air-/fuel ratio during combustion and the initial kinetic and thermodynamic conditions for auto ignition change. Figure 3-97 shows the achieved ignition angle at knock limit and the resulting time of fifty percent mass fuel burned as a function of EGR for engine operation at constant exhaust temperature pre turbine T_3 and engine load. Again a continous increase of ignition angle with increasing EGR rate can be observed. Next to this no definite difference between engine operation with and without EGR catalyst can be found at lower EGR rate. Primarily beyond six percent EGR ignition time can be advanced of about one degree crank angle.

Regarding the time of fifty percent mass fuel burned in Figure 3-97, right diagram, a general tendency to later combustion can be found. Despite the earlier ignition the deceleration of heat release due to modifications of cylinder mixture seems to be dominant. In comparison, heat release takes place slightly earlier when using the EGR catalyst.

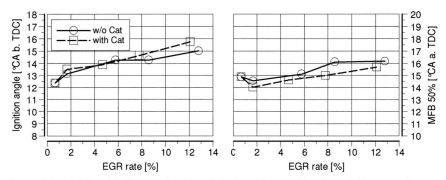

Figure 3-97: Achieved ignition angle at knock limit and the resulting time of fifty percent mass fuel burned (MFB 50%) as a function of EGR with and without EGR catalyst. $n = 3000\ min^{-1}$, $imep = 19\ bar$, $T_3 = 1173\ K$

The already discussed interactions of λ and EGR rate on initial kinetic and thermodynamic conditions at ignition time are displayed in Figure 3-98.

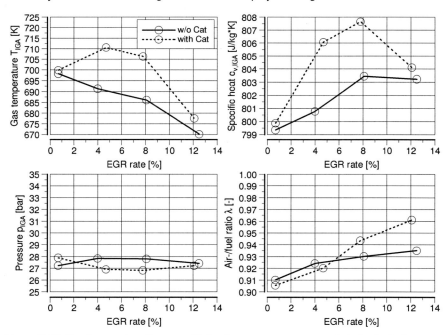

Figure 3-98: Gas temperature T_{IGA}, specific heat $c_{v,IGA}$, pressure p_{IGA} and air-/fuel ratio λ as a function of EGR with and without EGR catalyst. $n = 3000\ min^{-1}$, $imep = 19\ bar$, $T_3 = 1173\ K$

Contrary to stoichiometrical engine operation in Figure 3-95, T_{IGA} and the corresponding $c_{v,IGA}$ differ between both EGR modes. While T_{IGA} continuously decreases with increasing EGR rate in standard EGR mode, at low EGR rates a first increase with a successive decrease to higher EGR rates can be noticed with EGR catalyst. In regard to specific heat $c_{v,IGA}$, a linear increase up to eight percent is found corresponding to the increase of air-/fuel ratio without EGR catalyst. Using the EGR catalyst, an increase can be observed up to eight percent. Toward higher EGR rates, specific heat $c_{v,IGA}$ is quite equal to engine operation without EGR catalyst. At twelve percent EGR with quite similar cylinder pressure, a slightly increased temperature T_{IGA} in combination with a slightly increased specific heat $c_{v,IGA}$ is found, approving the dominant influence of air-/fuel ratio on caloric properties.

Finally, Figure 3-99 shows the calculated ignition delay time IDT with and without EGR catalyst in comparison to the achieved ignition angle. Whereas during engine run the achieved ignition angle does not differ much with tendency to more advanced ignition time using the EGR catalyst, IDT increases considerably stronger than without EGR catalyst. The more distinctive difference in the simulation approves the assumption of an overweighted λ influence on calculated IDT (comp. page 125).

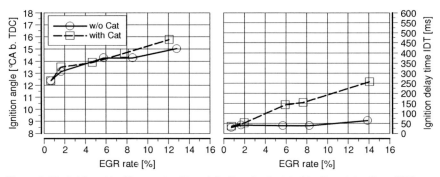

Figure 3-99: Achieved ignition angle at knock limit and calculated ignition delay time (IDT) as a function of EGR with and without EGR catalyst. $n = 3000\ min^{-1}$, imep = 19 bar, $T_3 = 1173\ K$

Summing up, engine measurements show no clear tendency of EGR catalyst effect on engine operation under enriched mixture condition and constant exhaust gas temperature. In general, the enleanment of the mixture with increasing EGR rate partially countervails the EGR effect. Additionally, conversion of EGR components like hydrocarbons and CO is reduced during catalytic conversion under enriched exhaust gas conditions. Although earlier ignition time is possible, time of heat released is delayed due to decelerated heat release. Nevertheless, fuel consumption is lowered due to decreased need for fuel enrichment. Using EGR catalyst shows slightly effects towards higher EGR rates. Here, increased engine efficiency can mainly be reasoned by a more isochoric heat release and consequently less loss due to incomplete combustion. Referring to this, no definite link to the use of the EGR catalyst is found. A definite reduced knock tendency with improved combustion time cannot be found. Simulation study shows qualitative trends of EGR effect on ignition delay time. The gradient of IDT versus EGR rate is lower compared to stoichiometrical case because of influences of enleanment of the mixture on IDT. Although based on test bench data, a definite increase of IDT can be found using the

EGR catalyst in simulations contrary to engine measurements.

A sensitivity analysis of EGR composition on auto-ignition tendency

A closer look at the numerical studies [221] gives informations about sensitivity of auto ignition towards the individual species found in the EGR. Numerical studies are initiated by species concentration measurements in the EGR route. For sensitivities study, each considered species is selectively reduced. Thus, mixture changes are corrected by adding or reducing nitrogen ratio. Figure 3-100 shows the delta of ignition delay time ΔIDT for each selectively reduced species and their conversion rate for an EGR rate of eight percent. The zero point is defined as the IDT when considering EGR composition in front of the EGR catalyst. The last column displays the ΔIDT when considering EGR composition behind EGR catalyst.

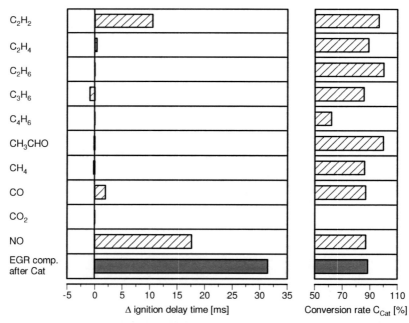

Figure 3-100: Sensitivity analysis of considered EGR species regarding their influence on ignition delay time and their conversion rate C_{cat} at eight percent EGR. $n = 2500\ min^{-1}$, $imep = 17.5\ bar$, $\lambda = 1$

IDT increases of nearly 10 ms due to the reduction of C_2H_2. C_2H_2 is an alkyne with a, in comparison to saturated hydrocarbon, high knock tendency [31]. It reacts via [227]

$$C_2H_2 + O_2 \rightarrow \dot{O}H + HCCO$$

$$(3\text{-}16)$$

with O_2 to the reactive hydroxyl (OH) radical. Hence, a reduction of C_2H_2 reduces OH radical production and therewith auto ignition tendency.

The concentration of the EGR component ethylene (C_2H_4) is reduced through catalytically conversion of about eighty-nine percent. This reduction slightly influence the ignition delay time of the mixture. C_2H_4 is known to have no strong knock tendency [31]. C_2H_6 shows a relatively low knock tendency [31]. Through catalytic conversion, the low ethane (C_2H_6) concentration is decreased down to detection limit. This reduction does not significant influence the ignition delay of the mixtures. An increase of this component should result in longer ignition delay times. The reduction of Propylene (C_3H_6) decreases the ignition delay time of the mixture. C_3H_6 has a RON about 90 [203]. Consequently, a reduction of C_3H_6 in a mixture with RON greater than 90 leads to a decrease of ignition delay time. On the other hand, reductions of a mixture with RON smaller than 90 leads to an increase of ignition delay time. It can be assumed, that the entire in-cylinder mixture has a RON smaller than 90. The 1,3-butadiene (C_4H_6) shows a concentration at the detection limit before catalyst and does not influence the IDT.

Acetaldehyde (CH_3CHO) is reduced nearly compeletely in catalyst. Because of its absolute low concentration, there is no significant effect on IDT. The concentration of methane (CH_4) is decreased through catalytic conversion by about eighty-six percent. This reduction leads to a slightly decrease in ignition delay time of the mixture. CH_4 is a component with a low knock tendency [203]. Therefore, an increase in CH_4 concentration should lead to an enhancement in ignition delay time.

CO is decreased in EGR catalyst down to 0.08 %. A decreased CO concentration leads to an increase of ignition delay time. CO reacts next with O_2 to CO_2 via [227]

$$\dot{HO_2}+CO \rightarrow \dot{OH}+CO_2 \text{ and} \tag{3-17}$$

$$\dot{OH}+CO \rightarrow \dot{H}+CO_2 \quad . \tag{3-18}$$

Reaction (3-17) converts hydroperoxyl (HO_2) with low reactivity to OH with high reactivity. Hence, the effect of this reaction is comparable to the NO conversion reaction (3-9). CO as EGR component has therefore a similar effect as NO. Reaction (3-18) also converts CO to CO_2 without any chain branching effect. The influence of CO on the autoignition process is therefore weaker than the influence of NO.

According to the three-way-catalyst mechanism, CO_2 is produced in EGR catalyst. This enhancement leads to no significant influence on the ignition delay time. It must be pointed out, that CO_2 leads to an increase of the specific heat capacity of the mixture and thereby influences the laminar flame speed.

The biggest influence on the ignition delay time is shown by a decrease of NO concentration in the mixture. NO is reduced through catalytic conversion from 1969 ppm to 254 ppm. In consequence, the ignition delay times increase of about 17.5 ms. As already mentioned, NO reacts with HO_2 to the reactive OH radical and nitrogen dioxide (NO_2). An enhanced OH radical concentration leads to acceleration of oxidation processes and therefore to shorter ignition delay times. A reduction of NO in the mixture results in longer ignition delay times.

Finally, the discussed sensitivity analysis explains the increased ignition delay time in the kinetic maps when reformatting EGR (comp. Figure 3-101).

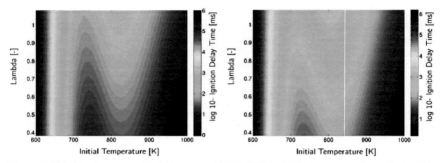

Figure 3-101: Kinetic maps with thirty percent EGR (left) and thirty percent reformatted EGR (right). $p = 20bar$, RON 95 (95% iso-octane / 5% n-heptane) [221]

The figures show a strong decrease in ignition delay times in the low temperature oxidation region, below 650 K, as well as in the high temperature oxidation region, beginning at 1100 K. In the region between the two main oxidation paths, also referred to as the Negative Temperature Coefficient (NTC) region, an enhancement of temperature leads to an increase of ignition delay time. Here, a temperature increase prevents chain abstraction and degeneration reactions and the degenerative chain branching reaction occurs as inverse of the reaction rate [227]. Starting at 730 K, the region dominates over a range of 90 K. The reason for this can be found in the enhanced reaction of NO and C_2H_2 to build OH radicals, which accelerates oxidation processes (compare Figure 3-100). By decreasing the concentrations of NO and C_2H_2, the oxidation process in the mixture in front of the propagating flame is slowed down. Considering the same velocity of propagating flame and compression of mixtures as without EGR Cat, the ignition angle is additionally moved to earlier crank angle position. The combustion process moves consequently to earlier crank angle position and therefore to a higher efficiency position as discussed previously. [221]

Excess air versus EGR and reformatted EGR dilution

Looking at kinetic maps of hydrocarbon/oxygen mixtures, a continous increase of ignition delay time with increasing air-/fuel ratio can be observed. Due to dilution with excess oxygen, auto ignition is delayed. For comparing effectiveness of EGR or oxygen dilution on knock occurrence, Figure 3-102 displays the achieved ignition angle at knock limit and the resulting time of fifty percent mass fuel burned as a function of the EGR, i.e. dilution rate, for engine operation without, with EGR Cat and enleanment. The dilution rate is defined as rate of excess oxygen during lean operation. During enleanment of cylinder mixture, the achieved ignition angle stays quite constant, while towards higher dilution rates the time of fifty percent mass fuel burned moves to later times. For all EGR/dilution rates, MFB50% is later using lean engine operation in comparison to EGR operation.

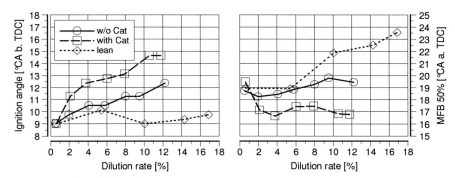

Figure 3-102: Achieved ignition angle at knock limit and time of fifty percent mass fuel burned as a function of EGR/dilution rate for engine operation with, without catalyst and enleanment. *n = 2500 min⁻¹, imep = 17.5 bar*

This is approved by the rates of heat release and burn functions for ten percent dilution in Figure 3-103.

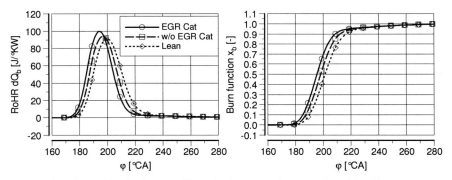

Figure 3-103: Rate of heat release and burn fuction for engine operation with, without catalyst and enleanment at an EGR/dilution rate of 10 percent. *n = 2500 min⁻¹, imep = 17.5 bar*

While the use of the EGR Cat results in earlier combustion time, heat release takes place distinctively later during lean operation. Next to this, the shape of heat release shows a tendentially slower heat release resulting in lower gradient of the burn function. The later combustion results in increased cyclic fluctuations like can be found in Figure 3-104 looking at the standard deviation of indicated mean pressure. The cyclic variations increase with increasing dilution rate and are higher than using EGR for dilution. In the left diagram of Figure 3-104 the indicated engine efficiency is shown.

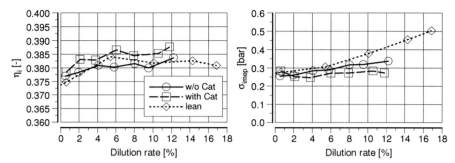

Figure 3-104: Indicated engine efficiency and standard deviation of indicated mean pressure as a function of EGR/dilution rate for engine operation with, without catalyst and enleanment. $n = 2500\ min^{-1}$, imep = 17.5 bar

Despite slightly lower engine efficiency at the starting point of variation, engine efficiency is tendentially higher than during EGR dilution without the EGR Cat considering measurement deviations of fuel weighting. The engine efficiency remains lower in comparison to operation with the EGR Cat. One characteristic difference between excess air and EGR dilution is the increased mass flow when using excess air, resulting in distinctively reduced specific emissions like can be found in Figure 3-105. With increasing dilution rate, HC and CO emission exponentially decrease until a dilution rate of fourteen percent and develop constant from there on. Due to increasing flame quench effects when diluting the cylinder mixture, HC concentrations increase. Considering the increased mass flow due to excess air, the specific HC emissions decrease. CO emissions decrease until minimum. Because of the excess air, there are enough oxidizers for composing CO_2 out of CO.

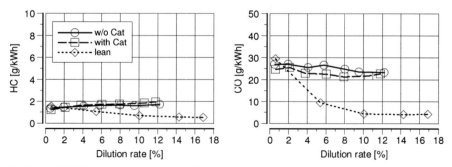

Figure 3-105: Unburned hydrocarbon and carbon monoxide emissions as a function of EGR/dilution rate for engine operation with, without catalyst and enleanment. $n = 2500\ min^{-1}$, imep = 17.5 bar

For the same reason, the NO_x emissions increase with increasing air dilution like can be observed in Figure 3-106. The maximum of NO_x emissions is reached at ten percent dilution rate equal to an air-/fuel ratio of $\lambda = 1.1$, which is kown to result in maximum NO_x emission [31]. The smoke emissions displayed in the right diagram of

Figure 3-106 slightly decrease when diluting with excess air approving the partial influence of air-/fuel ratio on oxidation of composed soot. Still carburetion artifacts seem to be the dominant reasons for soot emission in this engine operation condition.

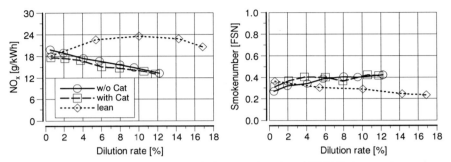

Figure 3-106: Nitric oxides and smoke emissions as a function of EGR/dilution rate for engine operation with, without catalyst and enleanment. *n = 2500 min⁻¹, imep = 17.5 bar*

Finally, comparing excess air and EGR dilution, excess air does not result in a recognizable reduction of knock tendency in this operation point. The continous retard of fifty percent mass fuel burned time as well as the calculated rate of heat release show a distinctive deceleration of heat release when diluting with excess air in comparison to EGR dilution. The resulting engine efficiency is quite comparable to the use of EGR dilution, while cyclic variations are considerably higher. Next to the increase of NO_x emissions, HC, CO and smoke emissions are lower when diluting with excess air.

4 Conclusion and forecast

The gasoline direct injection is a well known technology for improving SI-engines performance and efficiency. Today, the customer relevant efficiency benefits are limited by pumping losses during part-load and engine knock during full-load operation.

In the last two decades, the probability of stratifying the air-/fuel mixture using several combustion processes has been widely investigated for reducing pumping losses and improving part-load operation in direct injection SI-engines. With the spray-guided combustion processes, the second generation of gasoline direct injection has been put into production. Still, these are limited to an engine map area at low loads and engine speed. Using turbo charging, these limits can be effectively extended. On the other hand, stratified mixture formation with lean engine operation is limited by pollutant emissions. While NO_x emissions are widely discussed and understood, increased soot emissions during stratified engine operation needs to be taken into account. Next to the extension of the engine map area with stratified engine operation, turbo charging results in increased engine knock tendency at high- and full-load operation. External cooled exhaust gas recirculation (EGR) is a widely discussed approach to reduce knock tendency. While the thermodynamic effect of EGR is an accepted engine knock reducing effect, kinetic effects of recirculating burned, partially burned and unburned exhaust gas components are mostly neglected.

Combustion characteristics of stratified engine operation

Combustion of a spray-guided combustion process is mainly characterized by its spatial carburetion. The ring nozzle generates a ring of air-/fuel mixture in the combustion chamber, which is ignited at the exhaust side by the spark plug. The primary flame kernel drift toward the exhaust is reasoned by the present mixture. From there on, the primary flame front propagates through the combustion chamber. Non-premixed combustion phenomena occur behind the primary flame front. Non-premixed combustion evolves from heterogeneities of the stratified mixture formation and interaction of liquid fuel with the flame, for example occurring due to post injections for inflammation stabilization. Soot formation and oxidization are kinetically controlled processes due to their increased timescales. Thus, flame temperature in the zones of soot formation has to be investigated for describing soot formation. Using an imaging approach of the two-colour method, flame temperature distributions of non-premixed flames can be measured. The detected soot luminosity gives qualitative information about the formed soot concentration. In the context of test bench investigations and detailed thermodynamic analysis, the influences of engine parameters and fuel composition are demonstrated.

Optical combustion luminosity detection shows increased flame temperature, i.e. available combustion reactivity, in the area of the generated mixture ring towards the inlet. Below the injector in the center of the mixture ring, lower flame temperature can be observed (compare Figure 2-34). Corresponding CFD calculations verifies the dominant influence of the composed mixture ring on the occurrence of non-premixed combustion. The CFD calculated air-/fuel ratio distribution correlates with the measured soot luminosity as well as the resulting CFD calculated and optically determined temperature distribution during combustion. Simultaneous qualitative flame propagation analysis using an optical fibre spark plug sensor verifies the occurrence of non-premixed flame phenomena towards the inlet side. Distinctively increased light emissions can be observed. With closed tumble flap, a slight drift of the primary flame propagation and increased flame propagation speed can be observed. Decreasing exhaust gas air-/fuel ratio and fuel pressure results in a decreased flame propagation speed. Flame propagation becomes steadier during flame kernel formation. In opposition to that, an increased degree of homogenization caused by changing injection timings does not effect the primary flame propagation. Increasing EGR rate yields to a continous decrease of flame propagation speed. Hence, the spray-guided combustion process and the local occurrence of non-premixed combustion phenomena can be described as "carburetion controlled". The optical analysis improves the defintion of partially-premixed combustion for the combustion of stratified mixture formation in a DISI engine.

Determining the frequency distribution of non-premixed flame temperature, a normal frequency distribution is found. Comparing to the theoretical frequency distribution of soot particles as a function of flame temperature, the determined frequency distribution is shifted to higher temperatures. Regarding the determined non-premixed flame temperature, it has to be remarked, that, due to the physical approach of the measurement system, the surface temperature of the glowing particles is determined. Under theoretical considerations of temperature ranges of soot formation and oxidation, it can be assumed that this surface temperature is influenced by exothermic reactions during particle oxidation. This is verified by the only slight influence of mixture composition on non-premixed flame temperature in comparison to the zero-dimensional calculated temperature of the burned gas. In opposition to changes of the thermodynamic condition, the particle surface temperature is decreased when modifying the fuel composition by adding ethanol. Hence, the optically determined non-premixed flame temperature is driven by the thermal behavior of the formed soot particles and its surface reactions due to contemporaneous soot formation and oxidation.

Referring to test bench measurements, decreasing fuel rail pressure yields to increasing soot formation as well as increasing soot emissions. At low fuel pressure, soot luminosity remains for a longer period because of increased soot formation. The degree of homogenization is decreased due to increased fuel droplet sizes. A decrease of exhaust gas air-/fuel ratio leads to increased soot emissions. Below an air-/fuel ratio of $\lambda < 1.6$, soot emissions increase exponentially. Adding ethanol using dual-component fuel, even under global understoichiometrical conditions measurable soot emissions can be avoided. Soot oxidation is distinctively enhanced. Adding EGR, soot emisssions increase. Heat release is decelerated and shifted to the expansion stroke. Hence, gas temperature decreases and consequently less energy for soot oxidation is available. Using ethanol enriched dual-component fuels during EGR addition, cyclic variations increase. Running the engine with a synthetic multi-component ethanol enriched fuel, cyclic variations decrease as well as the EGR tolerance is increased. Leaning the local air-/fuel ratio at the spark plug by modifying the injection time, leads to distinctive decrease of soot emissions at higher engine load. Simultaneously, fuel consumption and NO_x emissions are decreased. Because of increased homogenization timescales, combustion mixture at the spark plug is leaned and the occurrence of non-premixed flame phenomena is reduced. This effect is limited by the exhaust gas air-/fuel ratio and the inflammation limit of the cylinder mixture. Using mono-/dual-component fuels, leaning limit is decreased to lower local air-/fuel ratios. Multi-component fuels show an increased leaning limit. Looking at the split of losses, ethanol addition leads to improved engine efficiency caused by an efficiency improving rate of heat release and reduced wall heat and chemical losses. These improvements are lowered by a more incomplete carburetion and combustion. A synthetic multi-component fuel shows similar efficiency improvements caused by its high degree of carburetion and more complete combustion like it has been indicated by its high EGR tolerance and increased leaning limit.

The influences of EGR composition on engine knock

Adding cooled EGR results in reduced knock tendency under high- and full-load engine operation. Because of the increasing specific heat (thermodynamic effect), gas temperature at the end of compression is decreased resulting in decreased auto ignition tendency. With the recirculation of exhaust gases, still reactive exhaust gas components are added to the cylinder mixture increasing its auto ignition tendency (kinetic effect). Using a three-way-catalyst, EGR is reformatted to CO_2 and H_2O at high-load engine operation. The measured emission concentrations up- and downstream of the EGR catalyst give the initial concentration for zero-dimensional kinetic simulation. A sensitivity analysis of the kinetic simulation gives information about auto ignition tendency of EGR components.

EGR leads to reduced knock tendency expressed by a shifted ignition time to earlier position. Heat release is shifted slightly to earlier time resulting in little decrease of fuel consumption. The effect of earlier ignition is partially countervailed by a decelerated heat release. Simulating the effect of EGR on auto ignition tendency, a qualitative similar behaviour of the calculated auto ignition delay time can be observed. Auto ignition delay time is increased as EGR rate increases. Using stochastic reactor networks, λ-T-maps of auto ignition delay time are generated. With increasing EGR rate, auto ignition delay time is reduced. EGR composition consisting of several unburned hydrocarbons as well as nitric oxide and carbon monoxide enhances knock tendency when considering identical λ-T-conditions. Thus, reduced knock tendency is exclusively reasoned by the decrease of compression end temperature. Reformatting EGR by EGR catalyst enables further knock tendency reduction due to the elimination of knock enhancing EGR components. Ignition time is shifted to earlier position. Heat release becomes more efficient resulting in a further decrease of fuel consumption of up to one percent. Split of losses approves the efficiency increasing effect of earlier heat release. Kinetical simulation shows a qualitative good correlation between achieved ignition angle and auto ignition delay time. Because of the EGR reformation, auto ignition delay time is distinctively increased reasoning the measured reduced knock tendency. Sensitivity analysis of kinetical simulation shows that ethyne C_2H_2, NO_x and CO mainly influence auto ignition delay time. The reduction of NO_x, C_2H_2 and CO leads to distinctively increased auto ignition delay time. The effect of reformatted EGR on knock tendency is reduced under enriched mixture condition. Because of an earlier ignition time and therewith reduced exhaust gas temperature, full-load enrichment is reduced resulting in leaner mixture. Leaning the mixture countervails the effect of EGR catalyst. Heat release is decelerated towards stoichiometrical air-/fuel ratio. Additonally, heat release accelerating hydrogen components in the EGR are reduced. Nevertheless, use of EGR at enriched high- and full-load engine operation yields to considerable decrease of fuel consumption.

Forecast

Although turbo charged spray-guided processes show a great theoretical potential for reducing fuel consumption. This potential cannot be used until today. The operation area on engine map with possible stratifiec engine operation is still limited to low engine loads and speeds by available exhaust gas aftertreatment systems. Ethanol enriched fuels enable the extension of this limit due to the simultaneous reduction of NO_x and soot emissions. The use of synthetic fuels can decrease possible soot emissions distinctively. Still particle emissions resulting from chamber and exhaust gas aftertreatment systems needs to be reduced for future emission legislations. Here, spray-guided combustion processes show benefits against wall-/air-guided combustion processes during engine cold start and engine heating periods.

Engineering high turbo charged and downsized SI-engines, an even greater potential for fuel consumption reduction under customer relevant engine operation can be found. Still these engines are limited by engine knock. For further efficiency improvement, high- and full-load EGR are essential. To gain higher improvements of EGR, combustion processes and engine periphery like turbo chargers need to be designed for EGR operation. Combustion enhancement by increased turbulent kinetic energy or the use of combustion enhancing components needs to be considered as well. Further approaches like oxygen depletion targets same effects.

5 Nomenclature

List of abbreviations

Abbreviation	Denotation
2-CRP	2-Colour ratio pyrometry
CAC	Charge air cooler
CAD	Computer assisted design
CAI	Controlled auto ignition
CARS	Coherent anti Raman spectroscopy
CCD	Charged couple device
CFD	Computional fluid dynamics
DISI	Direct injection spark ignition engine
ECU	Engine control unit
EGR	Exhaust gas recirculation
FID	Flame ionization detector
HCCI	Homogenous charge compression ignition
HD	Heavy duty
IRO	Intensified relay optic
LIF	Laser induced fluorescence
LII	Laser induced incandescence
LIS	Laser induced scattering
MBT	Maximum brake torque
MON	Motored octane number
NEDC	New European Driving Cycle
NSC	Nitric storage catalysator
NTC	Negative temperature coefficient
OEM	Original equipment manufacturer
PAH	Polycyclic aromatic hydrocarbons
PC	Personal computer / Pre chamber
PFR	Plug-flow reactor
RON	Research octane number
SC	Swirl chamber
SCR	Selective catalytic reduction
SI	Spark ignition

TDC		Top dead center (o = 180 °CA)
°CA		Crank angle degree
0-D DTDA		Zero-dimensional detailed thermodynamic analysis

List of formula symbols

Symbol	Unit	Denotation
A	m^2 / ms*bar^n	Surface / pre exponential factor
B	K	Factor proportional to energy level of activation
E_a	J/mol	Energy level of activation
E_B	-	Energy balance
F	-	Filter transmission
H	J	Enthalpy
H_G	J/g	Mixture heating value
H_u	J/g	Lower heating value
I_λ	W/m^2/m	Radiant density
IDT	s	Ignition delay time
K	-	Equilibrium constant
L	-	Air requirement
M	g/mol	Molar weight
P	- / kW	Products / Power
Q	J	Heat energy
R	J/g*K / -	Individual gas constant / Ratio
R_m	J/mol*K	Molar individual gas constant
S	counts	Signal
T	K	Temperature
U	J	Internal energy
V	m^3	Volume
Z	mA/W	Camera sensitivity
a	-	Linear coefficient / degree of conversion
b	-	Linear coefficient
c	J/g*K	Specific heat
h	J/g	Specific enthalpy
k	-	Reaction velocity constant

m	g / -	Mass / forming factor
n	-	Pressure exponent
p	bar	Pressure
s	m/s	velocity
t	s	Time
u	J/g	Specific internal energy
x	-	Fraction
α	-	Soot parameter
η	-	Degree of efficiency
ε	-	Degree of emission
λ	- / nm	Air-/fuel ratio / wavelength
μ	‰	Injection split factor
v'	m^2/s^2	Turbulent fluctuation velocity
σ	$W/m^{2*}K^4$	Boltzmann Constant
τ	s	Characteristic time of evaporation
ρ	g/m^3	Density
φ	°CA	Crank angle degree

List of indices

Indices	Denotation
B	Balance
F	Fuel
Cat	EGR catalyst
CS	Combustion start
CE	Combustion end
G	Mixture
IGA	Time of ignition angle
a	Apparent
air	Air
b	burn
bd	Burned gas
cyl	Cylinder

evap	Evaporation
exh	Exhaust
f	Flame
i	Indicated / injection / reaction index
l	Laminar
leck	Leckage
n	Index
puls	Pulsative
prop	Propagation
r	Residuals
s	Start / black radiator
s,rel	Soot luminosity relative
st	Stoichiometrical
ub	Unburned gas
v	Isochoric
vap	Vaporous
3	Before turbine

List of label abbreviations

Abbreviation	Unit	Denotation
C	%	Conversion rate
CO	g/kWh/ ppm	Specific carbon monoxide emissions / concentration
CO_2	g/kWh / ppm	Specific carbon dioxide emissions / concentration
EOI	°CA b. TDC	End of injection
HC	g/kWh / ppm	Specific unburned hydrocarbons emissions / concentration
IGA	°CA b. TDC	Time of ignition angle
ISEC	kJ/kWh	Indicated specific energy consumption
ISFC	g/kWh	Indicated specific fuel consumption
LI	[counts]	Light intensity
MFB 50%	°CA a. TDC	Time of fifty percent mass fuel burned
MFB 75%	°CA a. TDC	Time of seventy-five percent mass fuel burned
MFB 90%	°CA a. TDC	Time of ninety percent mass fuel burned
NO_x	g/kWh / ppm	Specific nitric oxides emissions

SOI	°CA b. TDC	Start of injection
$T_{avg.\ gas}$	K	Mass averaged gas temperature
$T_{burned\ gas}$	K	Burned gas temperature
T_{npf}	K	Non-premixed flame temperature

$dp/d\varphi$	bar/°CA	Maximum pressure gradient
imep	bar	indicated mean pressure
pmep	bar	pumping mean pressure
n_{mot}	min^{-1} / mol	Engine speed / quantity

σ_{Tnpf}	K	Standard deviation of non-premixed flame temperature
σ_{imep}	bar	Standard deviation of indicated mean pressure
ρ_{Tnpf}	-	Frequency distribution of non-premixed flame temperature
$\rho_{soot,normed}$	-	Normed frequency distribution of soot volume fraction

List of losses/efficiencies

Losses/efficiencies	Denotation
$\Delta\eta_\varepsilon$	Losses due to effective compression ratio
$\Delta\eta_\kappa$	Losses due to real cylinder charge
$\Delta\eta_{\alpha50\%}$	Losses due to isochoric heat release at MFB 50%
$\Delta\eta_{HC+CO}$	Losses due to incomplete combustion
$\Delta\eta_{RoHR}$	Losses due to finite heat release
$\Delta\eta_{Calorics}$	Losses due to changing gas properties
$\Delta\eta_{Wall\ heat}$	Losses due to wall heat transfer
$\Delta\eta_{Gas\ exchange}$	Losses due to gas exchange
$\Delta\eta_m$	Mechanical losses

η_e	Effective engine efficiency
η_i	Indicated engine efficiency
η_U	Engine conversion efficiency

6 References

[1]   Schintzel, K., Willand, J., Hoffmeyer, H., Schulz, R.: Neue ottomotorische Brennverfahren – Eine Alternative zum Dieselmotor. 7th Symposium „Direkteinspritzung im Ottomotor", HdT Essen, 2007

[2]   Fröhlich, K., Borgmann, K., Liebl, J.: Potenziale zukünftiger Verbrauchstechnologien. 24th Int. Symposium on engine technology Vienna, pp. 220-235, 2003

[3]   Dahl, D., Denbratt, I., Koopmanns, L.: An evaluation of different combustion strategies for SI engines in a multi-mode combustion engine. SAE-Paper 2008-01-0426, 2008

[4]   Lückert, P., Waltner, A., Rau, E., Vent, G., Schaupp, U.: Der neue V6-Ottomotor mit Direkteinspritzung von Mercedes-Benz. MTZ Vol. 67, pp. 830-840, 2006

[5]   Schwarz, C., Missy, S., Steyer, H., Durst, B., Schünemann, E., Kern, W., Witt, A.: Die neuen Vier- und Sechszylinder-Ottomotoren von BMW mit Schichtbrennverfahren. MTZ Vol. 68, pp. 332-341, 2007

[6]   Welter, A., Bruener, T., Unger, H., Hoyer, U., Brendel, U.: Der neue aufgeladene Reihensechszylinder-Ottomotor von BMW. MTZ Vol. 68, pp. 80-89, 2007

[7]   Lang, O., Habermann, K., Krebber-Hortmann, K., Sehr, A., Thewes, M.: Potenziale des strahlgeführten Brennverfahrens in Kombination mit Aufladung. Proceedings of "16th Aachener Kolloqium Fahrzeug- und Motorentechnik", 2007

[8]   Wirth, M., Zimmermann, D., Davies, M., Pingen, B., Borrmann, D.: Downsizing and stratified operation – an attractive combination based on a spray-guided combustion system. 27th Int. Symposium on engine technology, Vienna, pp. 68-85, 2006

[9]   Klüting, M., Missy, S., Schwarz, C.: Potentials of the spray-guided petrol direct injection combustion system in combination with turbo charging. 26th Int. Symposium on engine technology, Vienna, pp. 103-122, 2005

[10]   Fraidl, G.K., Kapus, P., Wiock, W.: Otto-Direkteinspritzung mit Aufladung – Die konkurrenz zu dieselmotorischen Antrieben? 26th Int. Symposium on engine technology Vienna, pp. 220-235, 2005

[11]   Kuberczyk, R.: Wirkungsgradoptimaler Ottomotor – Darstellung der Wirkungsgradunterschiede zwischen Otto- und Dieselmotoren. Bewertung wirkungsgradsteigernder Maßnahmen bei Ottomotoren. Final report FVV No. 875, 2008

[12]   Hoffmeyer, H., Willand, J.: Einfluss von Einspritzstrategien auf die Verbrennung von DI-Turbomotoren. 6th Symposium „Diesel- und Benzindirekteinspritzung", HdT Berlin, 2008

[13]   Leonhard, R., Bäuerle, M., Gerhardt, J.: Direkteinspritzung für saubere, sparsame und starke Ottomotoren. Technical Academy Esslingen (TAE), 2004

[14]   Merker, G.; Schwarz, C.;Stiesch, G.; Otto, F.: Verbrennungsmotoren – Simulation der Verbrennung und Schadstoffbildung. 2nd Edition; ISBN 3-519-16382-9 B.G.Teubner Stuttgart Leipzig Wiesbaden, 2004

[15]   Weberbauer, F., Rauscher, M., Kulzer, A., Knopf, M., Bargende, M. 2005. Allgemein gültige Verlustteilung für neue Brennverfahren. MTZ Vol. 66, pp. 120 – 124, 2005

[16]   Bargende, M.: Berechnung und Analyse innermotorischer Vorgänge. Lecture note University Stuttgart, 2004

[17]   Bargende, M., Burkhardt, C., Frommelt, A.: Besonderheiten der thermodynamischen

Analyse von DE-Ottomotoren. MTZ Vol. 62, pp. 56–68, 2001

[18]   Pischinger, R., Klell, M., Sams, T.: Thermodynamik der Verbrennungskraftmaschine. 2nd Edition, Springer Wien New York, ISBN 3-211-83679-9, 2002

[19]   Zacharias, F.: Analytische Darstellung der thermischen Eigenschaften von Verbrennungsgasen. Dissertation TU Berlin, 1966

[20]   McBride, B.J.: Thermodynamic properties to 6000 K for 210 substances involving the first 18 elements. NASA SP 3001, Washington D.C., 1963

[21]   Edson, M.H.: The influence of Compression Ratio and Dissociation on ideal Otto Cycle Engine efficiency. SAE Transaction 1962, pp. 665-679, 1962

[22]   De Jaegher, P.: Einfluss der Stoffeigenschaften der Verbrennungsgase auf die Motorprozessrechnung. Habilitation TU Graz, 1976

[23]   Grill, M.: Objektorientierte Prozessrechnung von Verbrennungsmotoren. Dissertation University of Stuttgart, 2006

[24]   Enginos GmbH: Thermodynamic analysis software Tiger. http://www.enginos.de

[25]   Woschni, G.: Die Berechnung der Wandverluste und der thermischen Belastung der Bauteile von Dieselmotoren. MTZ Vol. 31, pp. 491-499, 1970

[26]   Hoppe, N., Weberbauer, F., Zeilinger, K.: Modellbildung Otto-DI – Modellierungsverfahren für den Benzin-Direkteinspritzer-Verbrennungsprozeß. FVV Final report, 2001

[27]   Bargende, M., Briem, M., Greiner, R., Philipp, U.: Versuchs- und Meßtechnik an Motoren. Lecture notes University of Stuttgart, 2004

[28]   Weberbauer, F., Hoppe, N.: DVA-DI – Programm zur Analyse gemessener Druckverläufe von Direkteinspritzer-Ottomotoren. FVV Final report, 2001

[29]   Woschni, G., Fieger, J.: Experimentelle Bestimmung des örtlich gemittelten Wärmeübergangskoeffizienten im Ottomotor. MTZ Vol. 42, 1981

[30]   Koch, T.: Numerischer Beitrag zur Charakterisierung und Vorrausberechnung der Gemischbildung und Verbrennung in einem direkteingespritzten, strahlgeführten Ottomotor. Dissertation ETH Zürich (No. 14937), 2002

[31]   Heywood, J.B.: Internal combustion engine fundamentals. 2nd Edition, ISBN 0-07-100499-8, McGraw-Hill, 1988

[32]   Warnatz, J., Maas, U., Dibble, R.W.: Verbrennung. 3rd Edition, Springer Berlin Heidelberg New York, ISBN 3-540-42328-9, 2001

[33]   Arrhenius, S.: Über die Reaktionsgeschwindigkeit bei der Inversion von Rohrzucker durch Säuren. Journal of Physical Chemics, pp. 4-226, 1889

[34]   Glassman, I.: Combustion. Academic Press, San Diego, 1996

[35]   Semenov, N.N.: Chemical kinetics and chain reactions. Oxford University Press, London, 1935

[36]   Frank-Kamenetskii, D.A.: Diffusion and heat exchange in chemical kinetics. Princeton University Press, Princeton N.J., 1958

[37]   Douaud, A.M., Eyzat, P.: Four-octane-number method for predicting the anti-knock behaviour of fuels and engines. SAE-Paper 780080, 1978

[38]   Semenov, N.N.: Some problems in chemical kinetics and reactivity. Princeton University Press, Princeton, N.J., 1958

[39]   Bamford, C.H., Tipper, C.F.H.: Comprehensive chemical kinetics. Gas Phase Combustion Vol. 17, Elsevier Amsterdam Oxford New York, 1977

[40]   Metghalchi, M., Keck, J.C.: Laminar burning velocities of propane-air mixtures at high temperature and pressure. Combustion and Flame Vol. 38, pp.143-154, 1980

[41]   Zeldovich, Y.B., Frank-Kamenetskii, D.A.: The theory of thermal propagation of flames. Zh Fiz Khim, pp. 12-100, 1938

[42]   Peters, N.: Turbulent combustion. 4th printing, Cambridge University Press, ISBN 0-521-66082-3, 2006

[43]   Groff, E.G.; Matenukas, F.A.: The nature of turbulent flame propagation in a homogenous spark-ignited engine. SAE-Paper 800133, 1980

[44]   Linse, D., Hasse, C., Durst, B., Schwarz, C.: Kombinierte experimentelle und numerische Analyse der turbulenten Flammenausbreitung und –struktur in einem direkteinspritzenden Ottomotor. VDI-Reports No. 1988, 2007

[45]   Libby, P.A., Williams, F.A.: Turbulent reacting flows. Springer NewYork, 1980

[46]   Zimmermann, D., Kerek, Z., Wirth, m., Gansert, K.-P., Greszik, R., Storch, A., Josefsson, G., Sandquist, H.: Inflammation of stratified mixtures in spray guided DI gasoline engines: Optimization by application of high speed imaging techniques. 6th Int. Symposium on Combustion Diagnostics, Baden-Baden, pp. 107-122, 2006

[47]   Lippert, A.M., El Tahry, S.H., Huebler, M.S., Parrish, S.E., Inoue, H., Noyori, T., Nakama, K., Abe, T.: Development and optimization of a small-displacement Spark-Ignition Direct-Injection engine- stratified operation. SAE-Paper 2004-01-0033, 2004

[48]   Schintzel, K., Willand, J., Hoffmeyer, H.: Das strahlgeführte Brennverfahren im Spannungsfeld zwischen Potential und Kosten. 5th Symposium "Diesel- und Benzindirekteinspritzung", HdT Berlin, 2006

[49]   Lippert, A.M., Fansler, T.D., Drake, M.C., Solomon, A.S.: High-speed imaging and CFD modeling of sprays and combustion in a spray-guided spark ignition direct injection engine. 6th Int. Symposium on combustion diagnostics, Baden-Baden, pp.205-219, 2004

[50]   Kech, J.M., Reissing, J., Gindele, J., Spicher, U.: Analyses of the combustion process in a direct injection gasoline engine. COMODIA 98, pp. 287-292, 1998

[51]   Moriyoshi, Y., Komatsu, E., Morikawa, H.: Anaylsis of turbulent combustion in idealized stratified charge conditions. COMODIA 2001, pp. 226-231, 2001

[52]   Ando, H., Kuwahara, K.: A keynote on future combustion engines. SAE-Paper 2001-01-0248, 2001

[53]   Kuwahara, K., Ueda, K., Ando, H.: Mixing control strategy for engine performance improvement in a gasoline direct injection engine. SAE-Paper 940986, 1994

[54]   Schänzlin, K.: Experimenteller Beitrag zur Charakterisierung der Gemischbildung und Verbrennung in einem direkteingespritzten, strahlgeführten Ottomotor. Dissertation ETH Zürich, 2002

[55]   Spicher, U., Kölmel, A., Kubach, H., Töpfer, G.: Combustion in spark ignition engines with direct injection. SAE-Paper 2000-01-0649, 2000

[56]   Jakobs, Jan-Christian: Stickoxidbildung bei strahlgeführten Brennverfahren. Dissertation Otto-von-Guericke-University Magdeburg, 2009

[57]   Schintzel, K.: Kohlenwasserstoff-Emissionen eines Motors mit Benzin-Direkteinspritzung und wandgeführtem Brennverfahren. Dissertation Otto-von-

Guericke-University Magdeburg, 2005

[58] Sandquist, H.: Emission and deposit formation in direct injection stratified charge SI-engines. PhD Thesis Chalmers University Göteborg, 2001

[59] Willand, J., Schintzel, K., Hoffmeyer, H.: The potential of turbo charged spray-guided direct injection engines. MTZ worldwide edition Vol. 2, 2009

[60] Reissing, J.: Spektroskopische Untersuchung an einem Ottomotor mit Benzindirekteinspritzung. Dissertation University of Karlsruhe (TH), 1999

[61] http://eurlex.europa.eu/LexUriServ/site/de/oj/2007/l_171/l_17120070629de00010016.pdf

[62] Sato, H., Tree, D.R., Hodges, J.T., Foster, D.E.: A study on the effect of temperature on soot formation in a jet stirred reactor. Proceedings of the 23th International Symposium on Combustion, pp. 1469-1475, 1990

[63] Mauss, F.: Entwicklung eines kinetischen Modells der Rußbildung mit schneller Polymerisation. Dissertation RWTH Aachen, ISBN 3-89712-152-2 Cuvillier Verlag, Göttingen, 1998

[64] El-Gamal, M., Warnatz, J.: A computational study of soot formation in premixed combustion systems. 9th Tecflam Seminar, Soot and Carbon powder in flames, Stuttgart, 1993

[65] Frenklach, M., Wang, H.: Detailed mechanism and modeling of soot particle formation. In: Soot Formation in Combustion, Bockhorn, H. (Ed), pp. 162-193, ISBN 3-540-58398-X Springer-Verlag Berlin Heidelberg New York, 1994

[66] Chippett, S., Gray, W.A.: The size and optical properties of soot particles. Combustion and Flame Vol. 31, pp.149-159, 1978

[67] Smith, O.: Fundamentals of soot formation in flames with application to Diesel engine particulate emissions. Progress in Energy and Combustion Science Vol. 7, pp.275-291, 1981

[68] Wagner, H.G.: Soot formation in combustion. Proceedings of the Combustion Institute Vol. 17, 1979

[69] Wang, H., Frenklach M.: A detailed kinetic modeling study of aromatics formation in laminar premixed acetylene and ethylene flames. Combustion and Flame Vol. 110, pp. 173-221, 1997

[70] Hura, H.S., Glassmann, I.: Fuel oxygen effects on soot formation in counterflow diffusion flames. Combustion Science Technology Vol. 53, 1987

[71] Stein, S.E., Walker, J.A., Suryan, M.M., Fahr, A.: A new path to benzene in flames. Proceedings of the Combustion Institute, Vol. 23, pp.85, 1991

[72] Alkemade, V., Homann, K.H.: Formation of C6H6 isomers by recombination of propynyl in the system sodium vapour/propynykhalide. Journal of Physical Chemics, Vol. 161, pp.19, 1989

[73] Wang, H., Frenklach, M.: Detailed modeling of soot particle nucleation and growth. Proceedings of the 23rd International Symposium on Combustion, pp. 1559, 1990

[74] Haynes, B.S., Wagner, H.G.: Soot formation. Progress in Energy and Combustion Science Vol. 7, pp. 229-273, 1981

[75] Bockhorn, H.: Soot Formation in Combustion – Mechanisms and Models. ISBN 3-540-58398-X Springer-Verlag Berlin Heidelberg NewYork, 1994

[76] Bartenbach, B.: Untersuchung zur Rußbildung und Rußoxidation unter den

Bedingungen turbulenter Diffusionsflammen. ISBN 3-8265-4355-6 Shaker Verlag, Aachen (Germany), Dissertation University Karlsruhe (TH), 1998

[77]   Megaridis, C.M., Dobbins, R.A.: Soot aerosol dynamics in a laminar Ethylene Diffusion Flame. Proceedings of the 22nd International Symposium on Combustion, pp. 353-362, 1988

[78]   Mayer, K.: Pyrometrische Untersuchung der Verbrennung in Motoren mit Common-Rail-Direkteinspritzung mittels einer erweiterten Zwei-Farben-Methode. Dissertation University of Karlsruhe (TH), 2000

[79]   Charalampopoulos, T.T., Chang, H.: In Situ optical properties of soot particles in the wavelength range from 340 to 600 nm. Combustion Science and Technology Vol. 59, pp. 401-421, 1988

[80]   Kunte, S.: Untersuchungen zum Einfluß von Brennstoffstruktur und –sauerstoffgehalt auf die Rußbildung und –oxidation in laminaren Diffusionsflammen. Dissertation ETH Zürich (No. 15003), 2003

[81]   Furutani, H., Tsuge, S., Goto, S.: The dependence of Carbon/Hydrogen ratio on soot particle size. SAE-Paper 920689, 1992

[82]   Richter, H., Howard, J.B.: Formation of polycyclic aromatic hydrocarbons and their growth to soot – a review of chemical reaction pathway. Progress in Energy and Combustion Science Vol. 26, pp. 565-608, 2000

[83]   Haynes, B.S., Wagner, H.G.: Sooting structure in a laminar diffusion flame. Proceedings of "Deutsche Bunsengesellschaft für Phyikalische Chemie" Vol. 84, pp. 499-506, 1980

[84]   Bönig, M., Feldermann, C., Jander, H., Lüers, B., Rudolph, G., Wagner, H.G.: Soot formation in premixed C_2H_4 flat flames at elevated pressures. Proceedings of the 23rd International Symposium on Combustion, pp. 1581-1587, 1990

[85]   Kook, S., Bae, C., Miles, P., Choi, D., Picket, L. M.: The influence of charge dilution and injection timing on low-temperature diesel combustion and emission. SAE-Paper 2005-01-3837, 2005

[86]   Street, J.C., Thomas, A.: Carbon formation in premixed flames. Fuel Vol. 34.4, 1955

[87]   Hansen, J.: Untersuchung der Verbrennung und Rußbildung in einem Wirbelkammer-Dieselmotor mit Hilfe eines schnellen Gasentnahmeventils. Dissertation RWTH Aachen, 1989

[88]   Kubach, H., Mayer, K., Spicher, U.: Untersuchungen zur Realisierung einer rußarmen Verbrennung bei Benzindirekteinpritzung. Research Report FZKA-BWPLUS, 2001

[89]   Tree, D.R., Foster, D.E.: Optical soot particle size and number density measurements in a direct injection diesel engine. Combustion Science and Technology Vol. 95, pp. 313-331, 1994

[90]   Roth, P., Brandt, S., von Gersum, S.: High temperature oxidation of suspended soot particles verified by CO and CO_2 measurments. Proceedings of the 23rd International Symposium on Combustion, pp. 1485-1491, 1990

[91]   Kennedy, I.M.: Models of soot formation and oxidation. Progress in Energy and Combustion Science Vol. 23, pp. 95-132, 1997

[92]   Charkrabotry, B.B., Long, R.: The formation of soot and polycyclic aromatic hydrocarbons in diffusion flames – Part one. Combustion and Flame Vol. 12, pp. 226-236, 1968

[93]   Nagle,   J.,   Strickland-Constable,   R.F.:   Oxidation   of   carbon   between

1000-2000 °C. Proceedings of the 5th Carbon Conference Vol. 1, pp. 154-164, 1962

[94]   Park, C., Appleton, J.P.: Shock-tube measurements of soot oxidation rates. Combustion and Flame Vol. 20, pp. 369-379, 1967

[95]   Lee, K.B., Thring, M.W., Beer, J.M.: On the rate of combustion of soot in a laminar soot flame. Combustion and Flame Vol. 6, pp. 137-145, 1962

[96]   Fenimore, C.P., Jones, G.W.: Oxidation of Soot by hydroxyl radicals. Journal of Physical Chemics Vol. 71, pp. 593-597, 1967

[97]   Puri, R., Santoro, R.J., Smyth, K.C.: The oxidation of soot and carbon monoxide in hydrocarbon diffusion flames. Combustion and Flame Vol. 97, pp. 125-144, 1994

[98]   Schubiger, R.A.: Untersuchung zur Russbildung und −oxidation in der dieselmotorischen Verbrennung: Thermodynamische Kenngrößen, Verbrennungsanalyse und Mehrfarbenendoskopie. Dissertation ETH Zürich (No. 14445), 2001

[99]   Pischinger, F., Lepperhoff, G., Houben, M.: Soot formation and oxidation in diesel engines. In: Soot Formation in Combustion, Bockhorn, H. (Ed), pp. 382-395, ISBN 3-540-58398-X Springer-Verlag Berlin Heidelberg New York, 1994

[100]   Bensler, H., Bo, T., de Risi, A., Mauss, F., Montefrancesco, E., Netzell, K., Willand, J.: Prediction of non-premixed combustion and soot formation using an interactive flamelet approach. THIESEL Conference, 2006

[101]   Montefrancesco, E.: Prediction of non-premixed combustion and soot formation using an interactive flamelet approach. Dissertation Universita del Salento, 2007

[102]   Akihama, K., Takatori, Y., Inagaki, K., Sasaki, S., Dean, A.M.: Mechanism of the smokeless rich diesel combustion by reducing temperature. SAE-Paper 2001-01-0655, 2001

[103]   Mueller, C.J., Pitz, W.J., Pickett, L.M., Martin, G.C., Siebers, D.L., Westbrook, C.K.: Effects of oxygenates on soot processes in DI Diesel engines: Experiments and numerical simulations. SAE-Paper 2003-01-1791, 2003

[104]   Bockhorn, H., Schön, A., Streibel, T., Peters, N., Born, C., Antoni, C., Mayr, B., Winzer, R., Halfmann, J., Häntsche, J., Pittermann, R., Krümmling, N.: Kinetik der Rußentstehung und −oxidation in DI-Dieselmotoren bei Abgasrückführung. Proceedings of "8th Aachener Kolloqium Fahrzeug- und Motorentechnik", pp. 905-928, 1999

[105]   Bertola, A., Schubiger, R., Kasper, A., Matter, U., Forss, A.M., Mohr, M., Boulouchos, K., Lutz, T.: Characterization of Diesel particulate emissions in HD-DI-Diesel engines with common rail fuel injection: Influence of injection parameters and fuel composition. SAE-Paper 2001-01-3573, 2001

[106]   Dresen-Rausch, J.: Untersuchung der Verbrennung im direkteinspritzenden Dieselmotor mittels Vielfach-Lichtleiter-Messtechnik. Dissertation RWTH Aachen, 1991

[107]   Stevens, E., Steeper, R.: Piston wetting in an optical DISI engine: Fuel films, pool fires and soot generation. SAE-Paper 2001-01-1203, 2001

[108]   Witze, P.O., Green, R.M.: LIF and Flame-emission imaging of liwuid fuel films and pool fires in a SI engine during a simulated cold start. SAE-Paper 970866, 1997

[109]   Kittelson, D.B., Phipo, M.J., Ambs, J.L., Luo, L.: In-Cylinder measurements of soot production in a direct-injection diesel engine. SAE-Paper 880344, 1988

[110]   Matsui, Y., Kamimoto, T., Matsuoka, S.: Formation and oxidation processes of soot

particulates in a DI diesel engine – an experimental study via the two-colour method. SAE-Paper 820464, 1982

[111] Ogawa, H., Miyamoto, N., Yagi, M.: Chemical-kinetic analysis on PAH formation mechanisms of oxygenated fuels. SAE-Paper 2003-01-3190, 2003

[112] Stoner, M., Litzinger, T.: Effects of structure and boiling point of oxygenated blending compounds in reducing diesel emissions. SAE-Paper 1999-01-1475, 1999

[113] Nabi, N., Minami, M., Ogawa, H., Miyamoto, N.: Ultra low emission and high performance diesel combustion with highly oxygenated fuel. SAE-Paper 2000-01-0231, 2000

[114] Kajitani, S., Kajitani, S., Chen, C.L., Oguma, M., Alam, M., Rhee, K.T.: Direct injection diesel engine operated with propane-DME blended fuel. SAE-Paper 982536, 1998

[115] Akasaka, Y., Sakurai, Y.: Effects of oxygenated fuel and cetane improver on exhaust emission from heavy-duty DI diesel engines. SAE-Paper 942023, 1994

[116] Miyamoto, N., Ogawa, H., Nurun, N.Md., Obata, K., Arima, T.: Smokeless, low NOx, high thermal efficiency and low noise diesel combustion with oxygenated agents as main fuel. SAE-Paper 980506, 1998

[117] Franke, H.U., Tschöke, H.: Influence of fuels on the soot-particle geometry. Proceedings of the 2nd International Colloqium on Fuels, pp. 517-525, TA Esslingen, 1999

[118] Glyde, H.S.: Experiments to determine velocities of flame propagation in a side valve petrol engine. J. Inst. Petroleum Technologists, Vol. 16, pp. 756-776, 1930

[119] Withrow, L., Rassweiler, G.M.: Slow Motion shows knocking and non-knocking expositions. SAE-Transactions, Vol. 39, pp. 297-303, 1936

[120] Gaydon, A.G.: The spectroscopy of flames. 2nd edition, Chapman & Hall, London, 1974

[121] Tipler, P.A., Mosca, G.: Physik. 2nd German Edition, Elsevier GmbH, ISBN 3-8274-1164-5, 2006

[122] Docquier, N., Candel, S.: Combustion Control and Sensors. A review. Progress in Energy and Combustion Science Vol. 28, pp. 107-150, 2002

[123] Ohmstede, G.: Untersuchung der Verbrennungsabläufe im Ottomotor mittels kurzzeitspektroskopischer Methoden. Dissertation TU Braunschweig, 1994

[124] Abata, D.L., Myers, P.S., Uyehara, O.A.: Spectroscopic Investigations of Hydroxyl Radical Formation in the End Gases of a Spark Ignition Engine Utilizing a Dye Laser. SAE-Paper 780970, 1978

[125] Aust, V., Zimmermann, G., Manz, P.-W., Hentschel, W.: Crank-Angle resolved temperature in SI engines measured by emission-absorption spectroscopy. SAE-Paper 1999-01-3542,1999

[126] Achleitner, E.: Über den Zusammenhang zwischen Lichtemissionen und Energieumsatz bei Verbrennungsmotoren und ihre Anwendung zu deren Regelung. Dissertation TU Wien, 1984

[127] Planck, M.: Ueber eine Verbesserung der Wien'schen Spectralgleichung. "Verhandlungen der Deutschen physikalischen Gesellschaft" Vol. 2, No. 13, pp. 202 – 204, 1900

[128] Wien, W.: Über die Energieverteilung im Emissionsspektrum eines schwarzen Körpers. "Annalen der Physik", pp. 662-669, 1896

158

[129] Stefan, J.: Über die Beziehung zwischen der Wärmestrahlung und der Temperatur. "Sitzungsberichte der mathematisch-naturwissenschaftlichen Classe der kaiserlichen Akademie der Wissenschaften", Vol. 79, S. 391-428, 1879

[130] Boltzmann, L.: Ableitung des Stefan'schen Gesetzes, betreffend die Abhängigkeit der Wärmestrahlung von der Temperatur aus der electromagnetischen Lichttheorie. "Annalen der Physik und Chemie", Vol. 22, S. 291-294, 1884

[131] Zhao, H., Ladommatos, N.: Engine combustion instrumentation and diagnostics. ISBN 0-7680-0665-1, SAE International, 2001

[132] Kuhn, D.: Messung von Temperatur- und Konzentrationsprofilen mittels Laserinduzierter Fluoreszenz (LIF). Dissertation University Karlsruhe (TH), 2000

[133] Brackmann, C., Bood, J., Afzelius, M., Bengtsson, P.E.: Thermometry in internal combustion engines via dual-broadband rotational coherent anti-stokes Raman spectroscopy. Meas. Sci. Technol. Vol.15, pp. R13-R25, 2004

[134] Hasegawa, R., Sakata, I., Yanagihara, H., Johansson, B., Omrane, A., Alden, M.: Two-dimensional gas-phase temperature measurements using phosphor thermometry. Applied Physics B Vol. 88, pp. 291-296, 2007

[135] Reissing, J., Kech, J.M., Mayer, K., Girdele, J., Kubach, H., Spicher, U.: Optical Investigations of a gasoline direct injection engine. SAE Paper 1999-01-3688, 1999

[136] Sakurauchi, Y., Ryu, H., Iijima, T., Asanuma, T.: Combustion gas temperature in a prechamber spark ignition engine measured by infrared pyrometer. SAE Paper 870457, 1987

[137] McComiskey, T., Jiang, H., Qian, Y., Rhee, K.T., Kent, J.C.: High-Speed spectral infrared imaging of spark ignition engine combustion. SAE Paper 930865, 1993

[138] Aust, V., Zimmermann, G., Manz, P.-W., Hentschel, W.: Crank-angle resolved temperature in SI-engines measured by emission-absorption spectroscopy. SAE Paper 1999-01-3542, 1999

[139] Li, Y., Gupta, R.: Simultaneous measurement of absolute OH concentration, temperature and flow velocity in flame by photothermal deflection spectroscopy. Applied Physics B, Vol. 75, pp. 903-906, 2002

[140] Zhao, H., Ladommatos, N.: Optical diagnostics for soot and temperature measurement in Diesel engines. Progress in Energy and Combustion Science Vol.24, pp.221-255, 1998

[141] Geitlinger, H., Streibel, T., Suntz, R., Bockhorn, H.: Two-dimensional imaging of soot volume fractions, particle number densities and particle radii in laminar and turbulent diffusion flames. Proceedings of the 27th International Symposium on Combustion, pp. 1613-1621, 1998

[142] Dec, J.E., Espey, C.: Ignition and early soot formation in a DI Diesel engine using multiple 2-D imaging diagnostics. SAE-Paper 950456, 1995

[143] Kawamura, K., Saito, A., Yaegashi, T., Iwashita, Y.: Measurement of flame temperature distribution in engines by using a two-color high speed shutter TV camera system. SAE-Paper 890320, 1989

[144] Ning, M., Yuan-Xian, Z., Zhen-Huan, S., Hu-Guo-dong: Soot formation, oxidation and its mechanism in different combustion systems and smoke emission pattern in DI Diesel engines. SAE-Paper 910230, 1991

[145] Quoc, H.X., Vignon, J.M., Brun, M.: A new approach of the two-color method for determining local instantaneous soot concentration and temperature in a D.I. Diesel

combustion chamber. SAE-Paper 910736, 1991

[146] Mohammad, I.S., Borman, G.L.: Measurement of soot and flame temperature along three directions in the cylinder of a direct injection diesel. SAE-Paper 910728, 1991

[147] Zhang, L., Minami, T., Takatsuki, T., Yokota, K.: An analysis of the combustion of a DI Diesel engine by photograph processing. SAE-Paper 930594, 1993

[148] Li, X., Wallace, J.S.: In-Cylinder measurement of temperature and soot concentration using the two-color method. SAE-Paper 950848, 1995

[149] Arcoumanis, C., Bae, C., Nagwaney, A., Whitelaw, J.H.: Effect of EGR on combustion development in a 1.9l DI Diesel optical engine. SAE-Paper 950850, 1995

[150] Iida, N. Nakamura, M., Ohashi, H.: Study of Diesel spray combustion in an ambient gas containing hydrocarbon using a rapid compression machine. SAE-Paper 970899, 1997

[151] Hampson, G.J., Reitz, R.D.: Two-color imagning of in-cylinder soot concentration and temperature in a heavy-duty DI Diesel engine with comparison to multidimensional modeling for single and split injections. SAE-Paper 980524, 1998

[152] Bakenhus, M., Reitz, R.D.: Two-color combustion visualization of single and split injections in a single-cylinder heavy-duty DI Diesel engine using an endoscope-based imaging system. SAE-Paper 1999-01-1112, 1999

[153] Larsson, A.: Optical studies in a DI Diesel engine. SAE-Paper 1999-01-3650, 1999

[154] Kweon, C.-B., Foster, D.E., Shibata, G.: The effects of oxygenate and gasoline-Diesel fuel blends on Diesel engine emissions. SAE-Paper 2000-01-1173

[155] Svensson, K.I., Mackrory, A.J., Richards, M.J., Tree, D.R.: Calibration of an RGB, CCD Camera and interpretation of its two-color images for KL and temperature. SAE-Paper 2005-01-0648, 2005

[156] Azetsu, A., Sato, Y., Wakisaka, Y.: Effects of aromatic components in fuel on flame temperature and soot formation in intermittent spray combustion. SAE-Paper 2003-01-1913, 2003

[157] Wyszynski, L.P., Aboagye, R., Stone, R., Kalghatgi, G.: Combustion imaging and analysis in a gasoline DI engine. SAE-Paper 2004-01-0045, 2004

[158] Stojkovic, B.D.: Development and application of a time resolved optical diagnostic for soot temperature and concentration in a spark-ignited direct-injection engine. Dissertation University of Michigan, 2003

[159] Ma, H.; Stevens, R.; Stone, R.: In-Cylinder Temperature Estimation from an optical spray-guided DISI Engine with Color-Ratio Pyrometry (CRP). SAE-Paper 2006-01-1198, 2006

[160] Ma, H., Marshall, S., Stevens, R., Stone, R.: Full-bore crank-angle resolved imaging of combustion in a four-stroke gasoline direct injection engine. Proceedings of the Institute of Mechanical Engineering Vol. 221, pp. 1305-1320, 2007

[161] Naccarato, F.: Development of a two-color based experimental technique for soot concentration measurement. PhD Thesis Universita del Salento, Lecce, 2007

[162] Khúla, M.: Abbildende Zweifarbenpyrometrie als Mittel zur Brennraumuntersuchung. Diploma thesis STU Bratislava, 2006

[163] Bohren, C.F.; Huffman, D.R.: Absorption and scattering of light by small particles. John Wiley and Sons Inc., ISBN-13: 978-0471057727, 1983

[164] Hottel, E.G.; Broughton, F.P.: Determination of true temperature and total radiation

from luminous gas flames – Use of special two-colour optical pyrometer. Industrial and engineering chemistry, Vol. 4, No.2, pp. 166-175, 1932

[165] Gaydon, A., Wolfhard, H.: Flames. Chapman and Hall, London, 1970

[166] Hartley, R., Zissermann, A.: Multiple View Geometry in Computer Vision. ISBN 0521623049, Cambridge University Press, 2000

[167] N.N.: Pyrometry Package. DaVis Software Manual, LaVision, Göttingen, 2006

[168] di Stasio, S., Massoli, P.: Influence of the soot property uncertainities in temperature and volume-fraction measurements by two-colour pyrometry. Meas. Sci. Technol. Vol. 5, pp.1453-1465, 1994

[169] Hentschel, W., Ohmstede, G., Dankers, S., Kallmeyer, F., Schintzel, K.: Micro-invasive optical diagnostics for combustion process development of the Volkswagen FSI engine. FISITA F2008-12-058, 2008

[170] Winkelhofer, E.: Flammenmeßtechnik für Motorenentwickler – Visiolution Anwenderhandbuch Ottomotoren. AVL List GmbH, 2007

[171] Stojkovic, B.D., Fansler, T.D., Drake, M.C., Sick, V.: High-speed imaging of OH* and soot temperature and concentration in a stratified-charge direct-injection gasoline engine. Proc. of the Combustion Institute, Vol. 30, pp. 2657-2665, 2005

[172] Müller, A., Wittig, S.: Experimental study on the influence of pressure on sot formation in a shock tube. In: Soot Formation in Combustion, Bockhorn, H. (Ed), pp. 350-370, ISBN 3-540-58398-X Springer-Verlag Berlin Heidelberg New York, 1994

[173] Grzeszik, R., Arndt, S., Raimann, J., Ruthenberg, I., Seibel, C.: The effect of spray momentum and vaporization rate on mixture formation and combustion for spray-guided GDI combustion processes. 6th Int. Symposium on Combustion Diagnostics, Baden-Baden, pp. 189-202, 2006

[174] Vibe, I.I.: Brennverlauf und Kreisprozeß von Verbrennungsmotoren. VEB Verlag Technik, Berlin, 1970

[175] Bäcker, H., Tichy, M., Achleitner, E., Pischinger, S., Lang, O., Habermann, K., Krebber-Hortmann, K.: Untersuchung des Ethanolmischkraftstoffs E85 im geschichteten und homogenen Magerbetrieb mit piezoaktuierter A-Düse. 6th Symposium "Direkteinspritzung im Ottomotor" HdT Essen, 2007

[176] Turner, J.W.G., Pearson, R.J., Holland, B., Peck, R.: Alcohol-based fuels in high performance engines. SAE-Paper 2007-01-0056, 2007

[177] Celik, M.B.: Experimental determination of suitable ethanol-gasoline blend rate at high compression ratio for gasoline engine. Appl. Therm. Eng. Vol. 28, pp. 396-404, 2008

[178] Kapus, P.E., Fuerhapter, A., Fuchs, H., Fraidl, G.K.: Ethanol direct injection on turbocharged SI engines – Potential and challenges. SAE-Paper 2007-01-1408, 2007

[179] Nakata, K., Utsumi, S., Ota, A., Kawatake, K., Kawai, T., Tsunooka, T.: The effect of ethanol fuel on a spark ignition engine. SAE-Paper 2006-01-3380, 2006

[180] Taniguchi, S., Yoshida, K., Tsukasaki, Y.: Feasibility study of ethanol applications to a direct injection gasoline engine. SAE-Paper 2007-01-2037, 2007

[181] Aleiferis, P.G., Malcolm, J.S., Todd, A.R., Cairns, A., Hoffmann, H.: An optical study of spray development and combustion of ethanol, iso-octane and gasoline blends in a DISI engine. SAE-Paper 2008-01-0073, 2008

[182] Sandquist, H., Karlsson, M., Denbratt, I.: Influence of ethanol content in gasoline on speciated emissions from a direct injection stratified charge SI engine. SAE-Paper

2001-01-1206, 2001

[183] Smith, J.D., Sick, V.: The prospects of using alcohol-based fuels in stratified-charge spark-ignition engines. SAE-Paper 2007-01-4034, 2007

[184] Cahill, G.F.: DIGAFA – Direct Injection Gasoline and Fuel Adaptation. Technical Report EUCAR, 2005

[185] Zeldovich, Y.B.: Regime Classification of an Exothermic Reaction with Non-uniformal Initial Conditions. Combustion and Flame Vol. 39, pp.211-214, 1980

[186] Chevalier, C., Warnatz, J., Melenk, H.: Automatic generation of reaction mechanisms for description of oxidation of higher hydrocarbons. Proceedings of "Deutsche Bunsengesellschaft für Phyikalische Chemie" Vol. 94, pp. 1362, 1990

[187] Pöschl, M.: Einfluss von Temperaturinhomogenitäten auf den Reaktionsablauf bei der klopfenden Verbrennung. Dissertation TU Munich, 2006

[188] Schintzel, K, Willand, J.: Klopfphänomene am Beispiel ausgewählter Motoren der Volkswagen AG. 2nd Symposium "Klopfregelung für Ottomotoren", Berlin, 2006

[189] Scholl, D., Davis, C., Russ, S., Barash, T.: The volume acoustic modes of spark-ignited internal combustion chambers. SAE-Paper 980893, 1998

[190] Dreizler, A., Maas, U.: Influence of the temperature distribution on the auto ignition in the end gas of Otto engines. COMODIA 98, pp. 197-202, 1998

[191] Gu, X.J., Emerson, D.R., Bradley, D.: Modes of reaction front propagation from hot spots. Combustion and Flame Vol. 133, pp. 63-74, 2003

[192] Goyal, G., Maas, U., Warnatz, J.: Simulation of the transition from deflagration to detonation. SAE-Paper 900026, 1990

[193] He, L., Law, C.K.: Geometrical effect on detonation initiation by a nonuniform hot pocket of reactive gas. Phys. Fluids Vol. 8(1), pp. 248-257, 1996

[194] Smilyanovski, V.: Ein numerisches Verfahren zur Berechnung schneller Vormischflammen und der Deflagrations-Detonations-Transition. Dissertation RWTH Aachen, 1996

[195] König, G., Maly, R.R., Bradley, D., Lau, A.K.C., Sheppard, C.G.W.: Role of exothermic centres on the initiation of knocking combustion. SAE-Paper 902316, 1990

[196] Itoh, T., Nakada, T., Takagi, Y.: Emission characteristics of OH and C2 Radicals under engine knocking. SAE Japan, Series B Vol. 38 No. 2, pp. 230-237, 1995

[197] Shoji, H., Shiino, K., Watanabe, H., Saima, A.: Simultaneous measurement of light absorption and emission of end gas during the knocking operation. SAE Japan, Vol. 15, pp. 109-116, 1994

[198] Shoji, H., Yoshida, K., Saima, A.: Radical behavior in preflame reactions under knocking operation. ASME Thermal Engineering Conference, Vol. 3, pp. 207-214, 1995

[199] Bäuerle, B., Behrendt, F., Kull, E., Nehse, M., Warnatz,J.: Experimental investigation and simulation of the formation of formaldehyde in the unburnt end-gas of an otto engine. COMODIA, pp.245-248, 1998

[200] Möhlmann, G.R.: Formaldehyde detection in air by laser-induced fluorescence. Applied spectroscopy Vol. 39, p.98ff, 1985

[201] Griffiths, J.F., Whitaker, B.J.: Thermokinetic interactions leading to knock during homogenous charge compression ignition. Combustion and Flame Vol. 131, pp. 386-399, 2002

[202] Lovell, W. G.: Knocking characteristics of hydrocarbons. Industrial Engineering Chemics Vol. 40, pp. 2388-2438, 1948

[203] Spausta, F.: Treibstoffe für Verbrennungsmotoren. Zweiter Band: Eigenschaften und Untersuchung der flüssigen Treibstoffe. Die gasförmigen Treibstoffe. Springer-Verlag, Wien, 1953

[204] Ganser, J.: Untersuchungen zum Einfluß der Brennraumströmung auf die klopfende Verbrennung. Dissertation RWTH Aachen, 1994

[205] Green, R.M., Cernansky, N.P., Pitz, W.J., Westbrook, C.K.: The role of low temperature chemistry in the autoignition of n-butane. SAE-Paper 872108, 1987

[206] Suzuki, T., Sakurai, Y.: Effect of hydrogen rich gas and gasoline mixed combustion on spark ignition engine. SAE-Paper 2006-01-3379, 2006

[207] Bunsen, E., Grote, A., Willand, J.: Vom variablen Ventiltrieb zum hocheffizienten Brennverfahren. MTZ Conference "Ladungswechsel im Verbrennungsmotor" Stuttgart, 2007

[208] Alger, T., Gingrich, J., Mangold, B.: The effect of hydrogen enrichment on EGR tolerance in spark ignited engines. SAE Paper 2007-01-0475, 2007

[209] Topinka, J.A., Gerty, M.D., Heywood, J.B., Keck, J.C.: Knock behavior of lean-burn, H_2 and CO enhanced, SI gasoline engine concept. SAE-Paper 2004-01-0975, 2004

[210] Quader, A., Kirwan, J., Greive, M.: Engine performance and emissions near the dilute limit with hydrogen enrichments using an on-board reforming strategy. SAE-Paper 2003-01-1356, 2003

[211] Tsolakis, A., Megaritis, A., Yap, D., Abu-Jrai, A.: Combustion characteristics and exhaust gas emissions of a diesel engine supplied with reformed EGR. SAE-Paper 2005-01-2087, 2005

[212] Grandin, B., Angström, H.-E.: Replacing Fuel Enrichment in a Turbo Charged SI Engine: Lean Burn or Cooled EGR. SAE-Paper 1999-01-3505, 1999

[213] Cairns, A., Blaxill, H., Irlam, G.: Exhaust Gas Recirculation for Improved Part and Full Load Fuel Economy in a Turbocharged Gasoline Engine. SAE-Paper 2006-01-0047, 2006

[214] Grandin, B., Denbratt, I., Bood, J., Brackmann, C., Bengtsson, P.-E., Gogan, A., Mauss, F., Sunden, B.: Heat release in the end-gas prior to knock in lean, rich and stoichiometric mixtures with and without EGR. SAE-Paper 2002-01-0239, 2002

[215] Grandin, B., Angström, H.-E., Stalhammar, P., Olofsson, E.: Knock suppression in a turbocharged SI engine by using cooled EGR. SAE-Paper 982476, 1998

[216] Alger, T., Chauvet, T., Dimitrova, Z.: Synergies between high EGR operation and GDI systems. SAE-Paper 2008-01-0134, 2008

[217] Fraidl, G.K., Kapus, P.E., Prevedel, K., Fürhapter, A.: GDI Turbo: Die nächsten Schritte. 28th Int. Symposium on engine technology, Vienna, 2007

[218] Prabhu, S. K., Li, H., Miller, D. L., Cernansky, N. P.: The Effect of Nitric Oxide on auto ignition of a Primary Reference Fuel Blend in a Motored Engine. SAE-Paper 932757, 1993

[219] Kawabata, Y., Sakonji, T., Amano, T.: The Effect of NO_x on Knock in Spark-ignition Engines. SAE-1999-01-0572, 1999

[220] Stenlaas, O., Gogan, A., Egnell, R., Sunden, B., Mauss, F.: The influence of nitric oxide on the occurence of auto ignition in the end gas of spark ignition engines. SAE-

Paper 2002-01-2699, 2002

[221] Hoffmeyer, H., Montefrancesco, E., Beck, L., Willand, J., Ziebart, F., Mauss, F.: CARE – CAtalytic Reformated Exhaust gases in turbocharged DISI-Engines. SAE Int. J. Fuels Lubr. Vol.2(1), pp. 139-148, 2009

[222] Baronick, J., Heller, B., Lach, G., Luf, H.: Modal Measurement of Raw Exhaust Volume and Mass Emissions by SESAM. SAE-Paper 980047, 1998

[223] http://www.avl.com

[224] Manual AVL IndiCom 1.5b, AVL List GmbH, 2006

[225] Mauss, F.: Matlab-Tool for the Postprocessing of DARS calculations, unpublished, 2008

[226] Mauss, F., Pasternak, M., Seidel, L., Ingle, A., Meyer, P.: Kraftstoffkennzahlen – Kinetic Mapping. FVV Pre project, 2008

[227] Ahmed, S.S., A detailed modeling study for primary reference fuels and fuel mixtures and their use in engineering applications. PhD Thesis Lund University, 2006

[228] http://www.kistler.com

[229] Thiemann, W.: Ausgewählte Kapitel der Dieselmotorentechnik. Lecture note University of Stuttgart, 2004

[230] http://www.de.sgs.com/de

[231] http://www.cambustion.co.uk/

[232] Brettschneider, J.: Berechnung des Luftverhältnisses λ von Luft-Kraftstoff-Gemischen und des Einflusses von Messfehlern auf λ. Bosch Technical Report Vol.6, 1979

[233] http://www.diganars.com/

[234] Ahmed, S.S., Moréac, G., Zeuch, T., Mauss, F.: Reduced Mechanism fort he Oxidation oft he Mixtures of n-Heptane and iso-Octane. Proceedings of the European Combustion Meeting, Louvain-la-Neuve, 2005

[235] http://grimec.com/

[236] Ohmstede, G., Alberti, P., Kapitza, L., Lange, F., Schmerbeck, S.: Investigation of the Properties of Alternative Fuels for Diesel-Engine Combustion. 8th Int. Symposium on combustion diagnostics, Baden-Baden, pp.285-296, 2008

[237] Schintzel, K., Willand, J., Hoffmeyer, H.: The influence of alternative fuels on the spray-guided gasoline direct injection system. 8th Int. Symposium on combustion diagnostics, Baden-Baden, pp.285-296, 2008

[238] Hoffmeyer, H., Willand, J.: Optical analysis of non-premixed phases of combustion of a fuel matrix in a turbo charged gasoline direct injection engine. 9th Symposium „Internal engine combustion", HdT München, 2009

[239] Winkelhofer, E., Beidl, C., Philipp, H., Piock, W.: Optische Verbrennungsdiagnostik mit einfach applizierbarer Sensorik. MTZ Vol. 62, pp. 644-651, 2001

7 Appendix

A Test engine specifications

Table A-7-1: Engine specifications

Specification	Unit	Value
Type	-	4 inline
Bore	mm	82.5
Stroke	mm	92.8
Conrod length	mm	144
Compression ratio	-	9.9
Inlet valve open (1mm valve lift)	°CA a. TDC	28
Inlet valve opening time	°CA	190
Exhaust valve open (1mm valve lift)	°CA b. BDC	8
Exhaust valve opening time	°CA	200
Cam phaser range	°CA	40
Turbo charger	-	BWTS K04 6.88
Spark plug	-	M12 top electrode
Injector spray diverging angle	-	90°, cone shape

B Test bench measurement setup

For test bench investigations, a standard test bench at Volkswagen AG group research is used. Test bench consists of a standard DC dynamo, which can be used in generating mode as well as motoring mode. For engine conditioning, two independent water circuits (charge air and engine cooling) are used, controlled by test bench automatisation system. Engine oil is cooled by a heat exchange, which is integrated into the engine cooling. Time averaged temperature measurement is performed by standard thermo elements, time averaged pressure measurement is carried out using slow absolute pressure sensors. Using an ABB Sensyflow®, fresh air mass flow is determined. Measurement routine is controlled by a test bench automation system controlling the routine as well as managing measurement data from the test bench system and the sub-systems.

Measurement setup and routine during optical measurements

Table B-7-2 shows the relevant engine parameter for the standard operation point during stratified part load operation. The engine is operated stationary. Modifications of engine parameters are noted.

Table B-7-2: Standard engine operation point during stratified part load operation

Parameter	Unit	Value
Engine speed	min⁻¹	1500
Indicated mean pressure	bar	4
Stand. fuel pressure	bar	190
Stand. cam phasing	°CA	0
Stand. end of injection	°CA b. TDC	21.4
Stand. ignition time	°CA b. TDC	26.6
Stand. throttle position	%	100
External EGR mass fraction	%	0
Engine coolant temperature	K	368
Charge air temperature	K	303
Oil temperature	K	≈ 368

Figure B-7-1 shows the engine test bench setup for optical investigation of stratified combustion. Additonally, positions of measurement/sample points are marked.

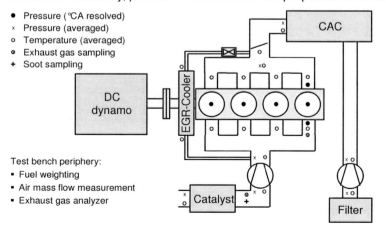

- ● Pressure (°CA resolved)
- × Pressure (averaged)
- ○ Temperature (averaged)
- ⊘ Exhaust gas sampling
- + Soot sampling

Test bench periphery:
- ▪ Fuel weighting
- ▪ Air mass flow measurement
- ▪ Exhaust gas analyzer

DC dynamo

EGR-Cooler

CAC

Catalyst

Filter

Figure B-7-1: Engine test bench setup for optical investigation of stratified combustion

Figure B-7-2 shows the measurement during optical investigations of stratified combustion. During optical investigations with the optical fibre spark plug sensor (Visiolution) and the 2-CRP system, exhaust gas emission concentrations are sampled at the exhaust port of cylinder one. This is noted at figure legends. For test bench investigations of fuel characteristics, exhaust gas emission concentrations are sampled in the turbine outlet. Indicated energy consumption and pressure based values are engine averaged.

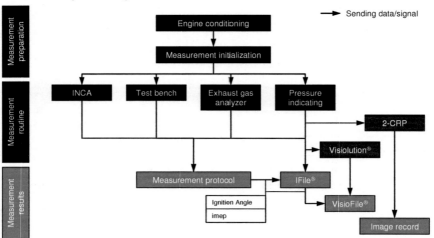

Figure B-7-2: Measurement routine for optical investigation of stratified combustion

Figure B-7-3 shows the injection and ignition strategy during stratified engine operation. The entire injection mass is controlled via the ECU. Using injection split factors the individual injection times are calculated.

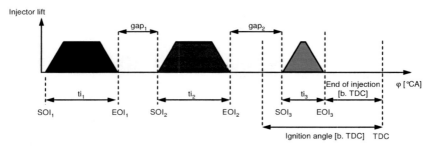

Figure B-7-3: Injection and ignition strategy during stratified engine operation.

Due to ignition stability, ignition angle and injection timing need to be constant. A change of the gap between ignition angle and injection timing results in misfire occurrence.

Measurement setup and routine during EGR catalyst investigation

Table B-7-3 displays the relevant engine parameter of the standard operations points during EGR catalyst investigations. The engine is operated stationary. Modifications of engine parameters are noted.

Table B-7-3: Standard operation points during high-load investigation

Parameter	Unit	Operation point 1	Operation point 2
Engine speed	min^{-1}	2500	3000
Indicated mean pressure	bar	17.5	19
Stand. fuel pressure	bar	150	150
Stand. cam phasing	°CA	-18	-23
Stand. start of injection	°CA b. TDC	305	305
Stand. ignition time	°CA b. TDC	9	12.4
Stand. throttle position	%	100	100
External EGR mass fraction	%	0	0
Stand. air-/fuel ratio	-	1	0.91
Lim. exhaust gas temperature	K	1173	1173
Engine coolant temperature	K	368	368
Charge air temperature	K	303	303
Oil temperature	K	375	378

Figure B-7-4 shows the engine test bench setup for EGR catalyst investigations. Additonally, positions of measurement/sample points are marked. Figure B-7-5 shows the measurement routine for EGR catalyst investigations. During measurement with EGR catalyst, three exhaust gas concentrations are sampled successively. For test bench analysis, specific emissions at the turbine outlet are calculated. EGR samples are used for catalyst conversion control and numerical investigations.

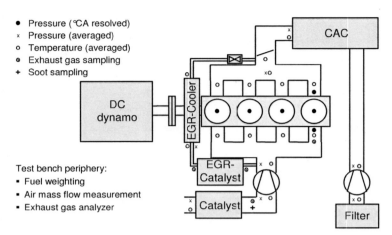

- Pressure (°CA resolved)
- Pressure (averaged)
- Temperature (averaged)
- Exhaust gas sampling
- Soot sampling

Test bench periphery:
- Fuel weighting
- Air mass flow measurement
- Exhaust gas analyzer

Figure B-7-4: Engine test bench setup for EGR investigation at high-load engine operation.

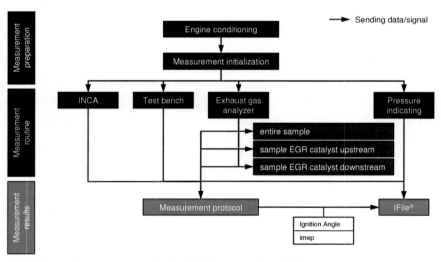

Figure B-7-5: Measurement routine for EGR investigation at high-load engine operation

C Pressure indicating

The test engine is equipped with piezo electric uncooled pressure transducers type AVL GU23D [223] in every cylinder and piezo resistive pressure transducers type

KISTLER 4075 A10 [228] in the intake and exhaust manifold runner of cylinder one. The exhaust pressure transduced is cooled using a water cooled shutter adapter type KISTLER 7531 [228], which closes the adapter during engine conditioning periods.

All pressure signals are amplified using AVL Microlfem® amplifier and recorded using an AVL Indimaster Advanced 671® with the indicating software AVL IndiCom 1.5b® [223]. During stratified engine operation, four hundred consecutive cycles are recorded. During EGR catalyst investigation, two hundred consecutive cycles are recorded. Relevant indicating values like indicated mean pressure *imep*, standard deviation of indicated mean pressure σ_{imep} and time of fifty percent mass fraction burnt *MFB 50%* are calculated online and saved in the measurement protocol.

During optical investigations, the indicating system triggers the optical measurement systems and sending relevant parameters like *record name*, *imep* and *ignition angle*. During EGR catalyst investigations, knock criterions of the indicating software (compare Figure C-7-6) is used for knock detection.

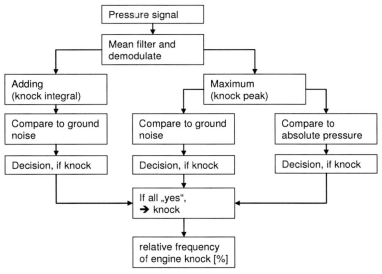

Figure C-7-6: Knock criterion of the pressure indicating software AVL IndiCom® 1.5b [223]

D Vibe-Parameter

For numerical calculations of internal combustion processes, a mathematical description of heat release is necessary. Several model categories have been developed, which can be classified regarding their degree of detail [16]. Merker et al. [14] classifies them into

- zero-dimensional or thermodynamic models,

- quasi-dimensional or phenomenological models and
- CFD models.

Vibe developed an approach for zero-dimensional description of heat release [174]. Based on reaction kinetic models, the normalized rate of heat release of a homogenous fuel/air mixture is described via an exponential function:

$$x_b = f(\varphi) = 1 - e^{-a \cdot z}$$ [A-1]

$$z = \frac{\varphi - \varphi_{CS}}{\varphi_{CE} - \varphi_{CS}}^{m+1}$$ [A-2]

with: a   degree of conversion (a = 6.91 corresponds to x_b = 0.999) and

     m   forming factor.

The degree of conversion a influences the final value of x_b and includes the losses due to incomplete combustion. The forming factor m determines the position of heat release maximum relative to the time interval of heat release. Thus, the forming factor and the time interval of heat release can qualitatively characterize the rate of heat release. A forming factor smaller than two leads to an asymmetrical shape with an fast increase in the beginning and a low gradient of decrease, a forming factor higher than two results in a low gradient of increase and a strong decrease on the falling edge of the shape. Typical Vibe parameters can be found for example at Thiemann [229]. Table D-7-4 summarizes characteristic Vibe-Parameter. It is found that the rate of heat release of diesel engines is characterized by forming factors smaller than one. In opposition to that, conventional gasoline engine heat release is described by a nearly symmetrical shape of heat release with forming factors higher than two.

Table D-7-4: Characteristic Vibe-Parameter according to Thiemann [229]

	Otto λ=1	Diesel SC	Diesel PC	Diesel DI	Diesel DI HD
Forming factor m	2.5 - 3	0.4 - 0.6	0.9 - 1.1	0.2 - 1 (TC)	0.2 - 0.4
φ_{CE}-φ_{CS} [°CA]	50 - 60	90 - 110	60 - 80	55 (TC) -89	70 - 80

     TC: Turbo charged

     SC: Swirl chamber

     PC: Pre-chamber

     HD: Heavy duty

E Test fuel specifications

Table E-7-5 shows test fuel specifications based on fuel data analysis [230]. Fuel data is used for test bench calculations like air-/fuel ratio determination as well as detailed thermodynamic analysis.

Table E-7-5: Test fuel specifications [230].

Specification	Label	Unit	Fuel Iso-Octane i-C8	E20%/ i-C8	E85%/ i-C8	DIGAFA	Ref-O_2-03 (OEC)
Research Octane Number	RON	-	100	99	144	99	100
Motor Octane Number	MON	-	100	97.3	96.2	85.3	88.1
Density		[kg/m^3]	0.6975	0.7139	0.7793	0.7307	0.755
Lower heating value	H_U	[MJ/kg]	44.45	39.84	28.83	41.76	43.129
Carbon	C	-	8	6.3342	2.5773	7.3	7.2
Hydrogen	H	-	18	14.9739	7.3467	15.0083	13
Oxygen	O	-	0	0.4794	1.0335	0.2041	0.01
Stoich. air-/fuel ratio	L_{st}	[-]	14.7172	13.7601	9.8196	14.2790723	14.4899443
Molar weight	M	[g/mol]	114.2656	98.8456	54.8964	106.015	99.5
Evaporation enthalpy	h_{vap}	kJ/kg	≈300	422.608	811.592	≈425	≈420
Vapor pressure	p_{vap}	kPa	52	34.2	28.5	54.1	60

Figure E-7-7 displays the boiling curves for the investigated fuels based on fuel data analysis.

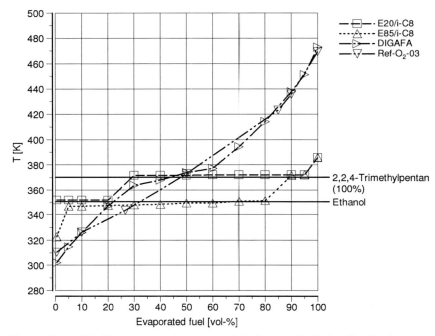

Figure E-7-7: Distillation curve for the used fuels and distillation line for iso-octane (2,2,4-Trimethylpentan) and ethanol. (According to DIN 31435)

F Pollutant emission of internal combustion engines

During internal combustion engine combustion, a hydrocarbon/oxygen mixture is converted ideally to CO_2 and H_2O. Since no ideal complete combustion occurs and a certain amount of nitrogen can be found in the supplied fresh air, CO, H, OH, O, NO and NO_2 as well as further radicals are composed during combustion. Thus, pollutant emissions are emitted, which can be seen as unhealthy. Figure F-7-8 shows the exhaust composition of a SIDI engine at stoichiometrical engine operation. These pollutant emissions can be classified in legally limited and non-limited emissions. Till today,

- carbon monoxide (CO),
- unburned hydrocarbons (HC) and
- nitric oxides (NO_x)

are legally limited in SI-engines. Formation mechanisms of limited emissions are widely investigated and summarized by Heywood [31] or Pischinger et al. [18] for example. In the future (Europe: EU5[+] and EU6 legislation [61]), particle emissions will be limited also. These particles are products of sooting combustion phenomena and deposit formation in the exhaust gas leading engine parts.

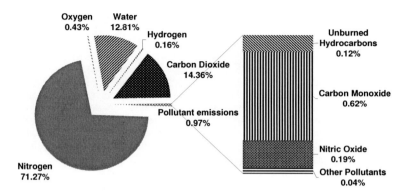

Figure F-7-8: Measured component concentrations in EGR at a EGR rate of 10 % EGR for operating conditions at *n = 2500 min⁻¹, imep = 17.5 bar, λ = 1* [221]

G Exhaust gas sampling and analysis

Exhaust emissions are sampled in the exhaust port of cylinder 1 for cylinder individual analysis (optical combustion characterization) and in the turbine outlet (test bench analysis) and analyzed using a SESAM3 exhaust emission measurement system [222]. Table G-7-6 shows the analyzed exhaust gas components using SESAM3 analysis. For unburned hydrocarbon measurement validation, exhaust gas is sampled in the turbine outlet and analyzed using a fast FID [231]. In front of the pre-catalyst, a sensor probe of the AVL SmokeMeter 415 S [223] is mounted for determining filter smoke number FSN. The EGR rate is determined using a carbon dioxide analyzer sampling the intake air and determining its carbon dioxide concentration. By this, the EGR rate can be calculated. For determining the exhaust gas air-/fuel ratio, it is calculated via the measured exhaust emission concentration using the method of Brettschneider et al. [232]. Considering fuel specifications, air mass flow is calculated.

Table G-7-6: Analyzed exhaust gas components using SESAM3 analysis.

CO	Carbon monoxide
CO_2	Carbon dioxide
NO	Nitric oxide
NO_2	Nitric dioxide
O_2	Oxygen
CH_2O	Formaldehyd
CH_4	Methane
C_2H_2	Acetylene
C_2H_4	Ethylene
C_2H_6	Ethane
C_3H_6	Propylene
C_3H_8	Propane (calc)
C_4H_6	1,3-Butadiene
AHC	Aromatic HC's
MECHO	Acetaldehyd
MEOH	Methanol
NMHC	NonMethanHC

H Modeling 0-D-Reactor-Network

Multiple reactor models, i.e. homogeneous reactors at constant volume and Plug Flow Reactors (PFR), are serially combined to obtain a reactor network able to replicate the combustion chamber and EGR pipe of the engine. In this way, the effect of EGR as also the influence of each EGR component on autoignition can be investigated. For this, the Software DARS [233] is employed together with a reduced iso-Octane/n-Heptane reaction mechanism [234]. The additional NO sub-mechanism is taken from GRI [235].

The zero-dimensional reactor network is displayed in Figure H-7-9. Input temperature, pressure, velocity and species concentration values are taken from test bed engine measurements in front of the EGR cooler and the mixture is successively cooled down to the temperature value measured behind the EGR cooler.

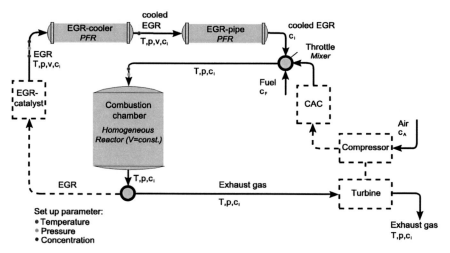

Figure H-7-9: Reactor-Network for numerical investigation of external EGR. [221]

The resulting temperature, pressure, velocity, and concentration data are fed into a second PFR, in which further heat losses occur. Afterwards, the cooled EGR is mixed with fresh air. Primary Reference Fuel with RON 95 (95 Volume-% iso-Octane and 5 Volume-% n-Heptane) is added to reach the target equivalence ratio. The fuel/air/EGR mixture is the input into a homogeneous reactor with constant volume. The initial temperature and pressure data are taken from heat release analysis data at ignition time. The system evolves until equilibrium conditions are reached. The ignition delay time is defined by the occurrence of the maximum of CO_2 concentration gradient. Beside the reactor network study, a *multi reactor system* is used. In this way it is possible to take into account temperature and equivalence ratio fluctuations existing in real engines. Using a multi reactor network system, the evolution of single reactors with different temperature and equivalence ratios can be considered. At which each single reactor evolves in time without any communication to the other reactors. The results of the multi reactor system calculations are presented in form of lambda-T-Plots [225] as found in [226].

I The endoscopic access

Figure I-7-10 shows technical scetches of the developed endoscope in a horizontal and a vertical intersection of cylinder one.

horizontal intersection   vertical intersection

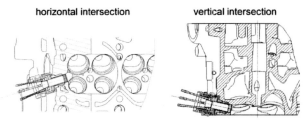

Figure I-7-10: Technical scetch of the endoscope developed for maximum light detection.

Figure I-7-11 displays a technical scetch as intersection of the three-sleeve endoscope.

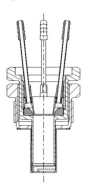

Figure I-7-11: Technical scetch of the three-sleeve endoscope with plan/concave saphir lense.

J The high pressure injection chamber

For basic investigations on penetration and evaporation characteristics of fuel injection a high pressure injection chamber is used. Figure J-7-12 shows the high pressure injection chamber used for these investigations. The chamber is equipped with three rectangulary arranged round windows for an optimal optical access. To avoid autoignition of the injected fuel the chamber is streamed continuously with nitrogen. The flow velocity is adjusted exactly to evacuate the chamber completely after injection. Generating a pressure range up to 100 bar and a gas temperature range up to 1273 K, injections during stratified mode can be simulated.

Figure J-7-12: High pressure injection chamber at Volkswagen Group Research. [236]

Finally, the injector is mounted perpendiculary on the fourth side of the chamber. Pressure, temperature and injector temperature are controlled using a standrad test bench control system. A serial production radial piston pump, standardly used in a diesel engine, generates the aspired fuel pressure. During the use of gasoline fuel, the pressure is limited to 1000 bar. Additionally, a lubrication additive is used for standard gasoline measurements.

Using a combined Mie-Schlieren technique, the penetration and evaporation characteristics of fuels injected by an outward opening nozzle can be described [1], [236]-[238]. With the combination of Mie-Scattering and Schlieren visualization, the liquid and the gaseous phase of the injected fuel can be qualitatively separated. Figure J-7-13 illustrates the setup of the combined Mie/Schlieren technology.

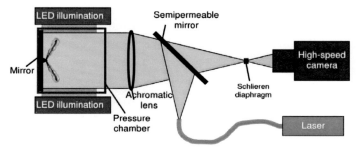

Figure J-7-13: Combined Mie/Schlieren setup in the injection pressure chamber [236]

Visualizing the liquid phase pulsed (t_{puls} = 10 ns) red light diode arrays (LED illumination) are used. The red light is scattered at the liquid fuel droplets and, using an achromatic lens, imaged on a high speed colour camera (repetition rate: 20 kHz). The gaseous phase is visualized using a schlieren setup. Via a semipermeable mirror and the acromatic lens, pulsed green laser light (t_{puls} = 200 ns) is inducted into the pressure chamber. A mirror, which reflects the parallel light, is mounted on the rear panel of the pressure chamber. The light beam is blocked using a schlieren aperture. Hence, only the refractive share of the reflected light is imaged on the camera chip. The optical refraction of the reflected laser light is caused by the density gradient generated by the fuel vapor. Finally, the gaseous and the liquid phase of the injected fuel can be imaged, separated qualitatively and recorded with a high time resolution of about 50 ns. According to Figure J-7-13 the camera view is in line with the injector axis. Hence, the radial fuel penetration can be imaged. The used piezo-driven fuel injector shows a circular fuel penetration during injection due to its outward opening ring nozzle. Next to the qualitative imaging evaluation of the fuel penetration and evaporation a numerical evaluation is performed. Because of the different colour of the liquid and the gas phase these can be seperated via binarization using a standard software algorithm in DaVis 7.1. With the known diameter of the mirror at the rear panel, the radial penetration of the liquid and the gaseous phase as a function of time can be determined. Comparing the progress of the penetrations, a qualitative evaporation rate can be defined.

K Technical Specifications of Two-ColourRatioPyrometry (2-CRP)

Figure K-7-14 shows the double imaging optic of the 2-CRP system.

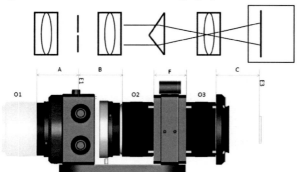

Figure K-7-14: Double imaging optic of the 2-CRP system. CAD image provided by LaVision GmbH

Table K-7 shows the technical specifications of the camera system.

Table K-7: Technical specifications of 2-CRP camera system.

Specification	Unit	Value
Camera model	-	IRO 18
CCD type	Pixel	Sony 1376x1040
Intensifier type	-	18mm V77090U-71-G131
Photocathode	-	GaAs
Phosphor	-	P43
Lens coupling	-	1.42 : 1
Effective pixel size	µm	9.2 @ Gate = 5 ns
Gating repetition rate	kHz	2

L The optical fibre spark plug sensor

For advanced combustion analysis, the AVL Visiolution® [239] system is used consisting of an optical signal recorder (OSR®) and the application software AVL VisioKnock® 1.2 beta-release. The optical fibre spark plug sensor is equipped with eighty optical channels detecting light intensity in a spectral range of $\lambda_w = 200 - 1000$ nm. For light detection, every channel is connected to a photo multiplier via optical fibre. The photo multiplier modulates the light signal into digital signals, which are evaluated as logical signals by the evaluation software. Figure L-7-15 shows a CAD image of the used optical spark plug sensor. The channels are classified into four views (35°, 45°, 67° and vertical view relative to horizontal engine plane). Because of this, the entire combustion chamber can be observed.

Figure L-7-15: CAD image of optical view of an eighty channel spark plug sensor. Provided by AVL.

The complete measurement routine is automatisated and controlled by the pressure indicating system giving cycle synchronized results corresponding to the recorded pressure traces.

8 Table of Figures

188